VOYAGING IN STRANGE SEAS

VOYAGING IN STRANGE SEAS

THE GREAT REVOLUTION IN SCIENCE

David Knight

YALE UNIVERSITY PRESS
NEW HAVEN AND LONDON

For information about this and other Yale University Press publications, please contact:
U.S. Office: sales.press@yale.edu   www.yalebooks.com
Europe Office: sales@yaleup.co.uk   www.yalebooks.co.uk

Set in Adobe Caslon Pro by IDSUK (DataConnection) Ltd
Printed in Great Britain by TJ International Ltd, Padstow, Cornwall

Library of Congress Cataloging-in-Publication Data

Knight, David M., author.
   Voyaging in strange seas: the great revolution in science/David Knight.
      pages cm
   ISBN 978-0-300-17379-6 (cl: alk. paper)
1. Science–History. 2. Discoveries in science–History. 3. Scientific expeditions–
History. I. Title.
   Q125.K578 2014
   509–dc23
                                                                2013041986
A catalogue record for this book is available from the British Library.

10 9 8 7 6 5 4 3 2 1

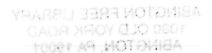

For Sarah, with thanks for fifty happy years;
and for Letsopa, Penelope, Laura and Hannah, beginning their lives

I could behold
The antechapel where the statue stood
Of Newton with his prism and silent face,
The marble index of a mind for ever
Voyaging through strange seas of Thought, alone.

William Wordsworth, *The Prelude* (1850)

The end of our foundation is the knowledge of causes, and secret motions of things; and the enlarging of the bounds of human empire, to the effecting of all things possible.

Francis Bacon, *New Atlantis* (1627)

CONTENTS

CHAPTER 1

Voyaging in Strange Seas

In humble and reflective mood, the elderly Sir Isaac Newton (1642–1727) said: 'I don't know what I may seem to the world, but as to myself, I seem to have been only like a boy playing on the sea-shore and diverting myself in now and then finding a smoother pebble or a prettier shell then ordinary, whilst the great ocean of truth lay all undiscovered before me.'[1] His admirer William Wordsworth (1770–1850) wrote of Newton embarking upon that ocean as he remembered the view from his college bedroom in Cambridge of

> The antechapel where the statue stood
> Of Newton with his prism and silent face,
> The marble index of a mind for ever
> Voyaging through strange seas of Thought, alone.[2]

It seems natural to us to see the scientific life as a voyage of discovery, with its highlights when we can say, like ancient mariners: 'We were the first that ever burst/ Into that silent sea.'[3] It became so because the long revolution that gave us modern science began with the ocean voyages of the 1480s and '90s. Vasco da Gama (c.1469–1525) sailed round the Cape of Good Hope into the Indian Ocean, opening up a new and more efficient route to the riches of India, China and the East Indies; and in 1492 Christopher Columbus (1451–1506) set out to reach them by going westwards, straight across the Atlantic Ocean. To his dying day, he believed that his landfall, the West Indies, was part of Asia. Discoverers often don't know quite what they have found; but the New World soon captured the imagination of Europeans, who began to think globally. Thomas More (1478–1535) set his Utopia on an island in the ocean, as did Francis Bacon (1561–1626), whose New Atlantis was run by an academy of sciences; the title-page of Bacon's *Instauratio Magna* (1620) shows ships sailing boldly through the Pillars of Hercules that bounded the Mediterranean into the open sea, and returning freighted with knowledge.

Nobody had suspected the existence of another continent, a brave new world to conquer literally and metaphorically. The Renaissance had seemed a recovery of the vast resources of ancient knowledge, and Aristotle (384–322 BC), Plato (427–347 BC), Ptolemy (c. AD 90–c. 168), Cicero (AD 106–43) and Pliny (AD 23–79) were adopted as authorities whose wisdom and style were to be emulated rather than questioned. Now their ignorance became apparent. When, in 1517, Martin Luther (1483–1546) launched what became the Reformation, he split the medieval Church, the other great source of legitimacy and certainty. Church teaching came to depend upon where you were. Nobody could any longer rely upon the old authorities. Luther translated the Bible into German, and William Tyndale (1494–1536) translated it into English, making the text available to be read and set against their own experience by those who were literate – a growing group. The Portuguese set up 'factories', trading stations and garrisons, in the East, and the Spaniards colonised Central and South America. The Dutch, French and British joined in. In 1620 the Pilgrim Fathers, Protestant refugees from England whose Church they thought insufficiently reformed, landed at Plymouth in what became in 1629 the Colony of Massachusetts; and it was here in 1775 that the first shots in the American War of Independence were fired. Our story ends in 1776 with the Declaration of Independence and the coming into being of the United States of America, whose first president, George Washington (1732–99), was inaugurated in 1789. Since that time, the course of science has moved gradually westwards, and the USA is now the leading scientific nation.

In the 1770s the project begun by the Portuguese and Spanish navigators of finding the limits of the habitable continents was essentially completed by James Cook (1728–79) in his three great voyages of discovery. The interiors of all except Europe remained unmapped – a task for succeeding generations – but their outlines were known. This image can also be applied to science. There had been a huge change in the way the world was perceived and understood, and agreement that it could be improved: life could be less arduous and more comfortable. A new empirical attitude prevailed – active, optimistic, based upon observation, experiment and mathematical reasoning. Curiosity that had seemed childish became a virtue. Science and its fruits were perceived by governments to be vital. This book traces this development, crucial to the making of our modern world, which applied inherited knowledge and practical techniques, new and rigorous (if provisional) standards of classification and explanation, new devices for observation and measurement, and entailed the transformation of experience into experiment. Maps began to indicate how to get to Rome or America, rather than to Heaven. Clockwork brought a new sense of time, a fascination with machinery, and a new vision of the heavens, animals and humans. Printing made accurate texts available even to people like Copernicus (1473–1543), living far from the centre of things in Poland, and no less crucially, made the suppression of knowledge more difficult.

Gentry and professional men (clergy, doctors and lawyers) began to devote time to what we call science, and to take an interest in what craftsmen called their 'mysteries' – trade secrets. Associated with courts, universities and printing houses, at first in Italy, then further north, they formed groups that coalesced into societies and academies. There had previously been soloists like Leonardo da Vinci (1452–1519), but now there was a chorus to carry science forward. It became a profitable business. Science and navigation required better instruments and thus promoted new trades; gentlemen of science took the time and trouble to find out what artisans were doing, and themselves began to think with their hands; chemists learned, and improved, techniques used by craftsmen in metallurgy and dyeing; and medical sciences were transformed as first-hand study of anatomy and physiology augmented clinical experience, though at first with little therapeutic improvement. New industries promoted science and trade, bringing empire, prosperity and leisure. Astronomers, botanists and zoologists marvelled at the order and variety in the world which they sought to classify; but God became for many a Clockmaker to be wondered at rather than prayed to.

The long Scientific Revolution had by the 1770s established its foundations, but this was not yet the modern world. We define 'science' narrowly; our forebears did not. They had no word for scientist, and there were no such people in the modern sense. They lived in a different world, foreign to us, where few people could devote themselves to science, but where there was a new spirit of curiosity about the natural world and how it worked – a spirit of progress. Projectors boldly schemed to improve the world, mitigating the biblical curse upon Adam and Eve by labour-saving devices, new crops, and better understanding of and treatments for disease. From the start, what we call science and technology were intertwined, inosculating and inseparable, and science was about power, wealth and usefulness as well as sociable and intellectual activity. But they remained unspecialised and, except for medicine and mathematics, there was no clear educational route into them: would-be practitioners relied on patronage, informal apprenticeships, and (most importantly) joining the societies and academies that began to proliferate.

For science cannot be done on its own, and its growth depended not only on its heroes (Galileo, Newton, Linnaeus) but also on the presidents, secretaries and treasurers of scientific societies, the craftsmen who made apparatus, the printer-publisher-booksellers who turned scientific discoveries into public knowledge, and the editors who invented the scientific journal: all of them interesting, if unsung, people. We have distorted the picture by too much concentration upon geniuses. This book emphasises, then, as crucial to the development of modern science its institutions (academies and societies) and its international character, promoted by travel and printing but also greatly by wars and persecution which led to the migration of well-educated refugees, notably to Britain and the Netherlands, and then later to the fledgling USA.

Many of these conflicts were so-called 'wars of religion', but in contrast to the popular idea of science as developing in warfare with religion, this book traces their complex engagement, with the mutual support they provided being as important to the outcome of the Scientific Revolution as the battles they fought. Certainly, the variety, order and beauty being revealed in the world increased wonder and respect for the Creator, who, however, seemed increasingly remote from a world that ran itself like a great clock. Science did not – and does not – answer all questions. To know what disease a person has, and what the prognosis is, is one thing; but the question 'why is this happening to me?', and how that person can best live with their condition, is quite another. And the eighteenth century saw not only the rise of Enlightenment Deism, with its remote First Cause, but also of Methodism and other pietist movements. On my reckoning, the Scientific Revolution coincides with what historians call the Long Reformation, from Luther to John Wesley (1703–91), with its emphasis upon private judgement and experience, unmediated by a priesthood. The phenomena are not unrelated.

The book sets chemistry, with its long tradition of laboratories, experimentation and useful knowledge, and natural history, with its classifications and its botanic gardens promoting agriculture and horticulture, at the heart of the Scientific Revolution – and considers medicine to be at least as important as astronomy and mechanics. In choosing voyages rather than publications as marking its limits, the book makes this period of 'revolution' or development even longer than other writers have done. But in so doing, it brings out the importance of seaborne empires (commercial as well as conquistador) in the rise of science, as well as of the new financial and statistical world that came into being with science. Because science is a social phenomenon, the book places it in contexts of war, wealth, crafts and industry, health and disease, publishing and communications, class, careers and leisure. It traces the slow build-up of broadly conceived science, subject to all sorts of contingencies and often poorly recognised and rewarded, becoming by the 1770s a key to understanding, power and wealth, unstoppable and ready for a real revolution in the Age of Revolutions.

For the chemist Humphry Davy (1778–1829), who was involved in that revolution, the great men of science, his predecessors, had all too often like Galileo (1564–1632) been

> despised or neglected; and great, indeed, must have been the pure and abstract pleasure resulting from the exercise of intellectual superiority and the discovery of truth, and the bestowing of benefits and blessings

upon society, which induced men to sacrifice all their common enjoy-
ments and all their privileges as citizens, to these exertions.[4]

For society, science was a road to riches, but not for all its practitioners
individually. Davy, for whom science had been a rapid route to social
mobility, was respected and honoured, and would be commemorated in
Westminster Abbey. But others were less fortunate, as we shall see.

Above all, this is a story about people. European males occupy the
foreground, because that it how it was, but they are not the whole cast. The
people we shall meet on this voyage – or strange, eventful pilgrimage – are
almost as varied in appearance, character and background as those whom
early travellers reported that they had encountered in the Americas, in
Africa and in the Indies. Institutions had their agendas, but because the
progress of science depended upon individuals and groups, and was subject
to the enthusiasms, whims and fortunes of powerful patrons, it was contingent: things could have been otherwise. This is a history with losers as well
as winners: astrologers like John Dee (1527–1608) found their science
passing out of favour, and skilled workmen were displaced by machinery
or by changes in practice. Travel and family history can stimulate our
historical imagination. The Second World War looks much less straightforward from Finland, Estonia, Romania or Hungary; and we may we find
not only rogues among our ancestors, but others who, hindsight tells us,
were on 'the wrong side': puritans, recusants, wheelwrights, farm labourers.
They did not feel part of a triumphant course of history. The history of
science is not all sweetness and light either; controversy is important in it,
and so is priority, though this may be disputed or wrongly assigned. There
are no silver and bronze medals in science: the perceived discoverer takes all
the credit. Hatred and malice are not excluded from our story: Newton
made enemies, and John Woodward (c.1665–1728), challenging another
doctor to a duel, said that he would rather face his sword than his physic.
Nor is sadness: in a fit of depression, Ralph Wood (d. 1726), who built the
first railway bridge (1725/6), threw himself from it to his death.

But there are also many attractive characters who demonstrate the
scientific virtues. Johannes Kepler (1561–1630) was certain that there must
be simple mathematical laws governing the planets and worked out a

brilliant geometrical model to demonstrate them. Hearing about Tycho Brahe's (1546–1601) exact observations, he sought the Dane out, only to find that his own treasured theory didn't fit Tycho's new data. Kepler then spent further years number-crunching until his new laws emerged to justify his faith in celestial harmony. The disciples of the great taxonomist Carl Linnaeus (1707–78) went out joyfully in search of new plants for their master to describe and classify, to distant places where many of them died as martyrs of science. Georg Wilhelm Richmann (1711–53), trying to repeat Benjamin Franklin's (1706–90) experiment, was electrocuted flying a kite on a thundery day. Others in our story suffered in health and in pocket through their devotion to science, unable, like Denis Papin (1647–1712), to find a steady job or, like Walter Charleton (1620–1707), to keep up their medical practice. Galileo was not the only one to be put in gaol; nor was Blaise Pascal (1623–62) unique in deserting science after undergoing religious conversion to a deeper and more demanding faith. Newton devoted much time to studying and interpreting the prophecies of Daniel and working out the ground plan of Solomon's Temple.

In a lighter vein, we shall meet Robert Burton (1577–1640) enthusing about tobacco as a panacea, better than an elixir, for melancholy; Johann Beringer (1667–1740) falling victim to a hoax when colleagues planted bogus fossils for him to find; and Robert Hooke (1635–1703) allowing a gnat to bite him before examining and drawing it under his microscope. Hooke also published his famous law relating to springs, 'ut tensio sic vis' ('as the extension, so the force'), as an anagram, 'ceiiinosssttuv', to puzzle contemporaries. Some whom we might have thought austere intellectuals devoted to abstract thought turn out to have been men of parts. Thus René Descartes (1596–1650) was first a soldier, joining the Protestant side and then the Catholic one in the hope of seeing action in the supposedly ideological Thirty Years' War (1618–48); afterwards he carried out optical experiments, and dissected the heart, illustrating his findings by means of pop-up flaps. The founders of modern science were diverse and bold, revolutionary indeed; they were also deeply rooted in their times.

Steven Shapin in 1998 opened his book *The Scientific Revolution* magnificently with the remark: 'There was no such thing as the Scientific Revolution and this is a book about it.'[5] Certainly, we expect revolutions to begin stirringly, with King Charles I raising his standard, Paul Revere's ride, the fall of the Bastille or the storming of the Winter Palace, and to end decisively after a few years, with Cromwell, Washington, Napoleon or Lenin firmly established in power. Imprecisions notwithstanding, it is still convenient to use the term the Scientific Revolution. It is a useful expression but Einstein gave us a clue to its proper signification when he referred to 'Fortunate Newton, happy childhood of science!'[6] We are faced with growth and development, not some explosive overturning of everything.

We pass through childhood and adolescence, and perhaps science grows and matures as we do. Adam and Eve are representative teenagers.[7] Falling in with the wrong company, impatient of orders and safety, wanting to be like gods, distinguishing between good and evil for themselves rather than merely by obedience to authority, they were ejected from the Garden of Eden as rebels. Aware that they were naked, shameful, self-conscious, passionate for knowledge, they were thrown out, going forth to increase and multiply, to till the soil by the sweat of their brows, to bear children in pain, and then to die. But after trying to pass the buck, they began to take responsibility for the voyage of their lives. John Milton's (1608–74) poetic epic *Paradise Lost* ends:

> The world was all before them, where to choose
> Their place of rest, and providence their guide;
> They hand in hand with wandering steps and slow,
> Through Eden took their solitary way.[8]

Growing up is a long and painful process but – despite nostalgia for the lost world of childhood – there are compensations. To diminish toil and pain and forestall death became tasks for humanity, in which science has played a crucial role. We might think of the sixteenth, seventeenth and eighteenth centuries as the adolescence of science. Not everyone, or every scientific start, survived that long; in Newton's time more died as babies or children than survived their teens. Science did survive: and the nineteenth century, following upon the Age of Revolutions, was the Age of Science, in which it matured to fulfil the promise shown earlier and came to dominate our culture. Human adolescence can go well or badly, but the comparison with science cannot be exact, for there was nothing necessary about its growth; it was contingent upon all sorts of circumstances. Most would-be revolutions fizzle out in riots, revolts and rebellions, crushed by the powers that be. It might have been so with science, but it wasn't. Science came through its adolescence in fine fettle: its story has become our story, and here it is.

CHAPTER 2

THE DEEP ROOTS OF MODERN SCIENCE

A THOUSAND YEARS SEPARATE THE decline and fall of the Roman Empire in the West from the fall of Constantinople in 1453. Later generations dubbed those years the Middle Ages, largely given over to superstition and ignorance, in the wake of the invasion of Vandals who smashed and grabbed, Huns and Vikings who raped and looted, and Goths who built gloomy and asymmetrical buildings upon the ruins of elegant classical edifices. To Victorian Darwinians, it even seemed as though medieval society had selected against civilised qualities: those who were more intelligent and peaceable went into monasteries and nunneries, leaving no descendants, while the unquestioning and the violent propagated their kind. The Renaissance seemed, as its name suggests, literally a rebirth, a time of youthful vigour and a fresh start. The Scientific Revolution marked a further leap in which the knowledge and achievements of the past would be exceeded: the Golden Age lay not in the past, but in the future.

After the break-up of the western Roman Empire, some small centres of learning, notably in Ireland, then in Northumbria and at Charlemagne's court, flourished, but much knowledge was lost in what used to be called the Dark Ages, down to about the millennium year 1000. Latin, the language of the Church, survived, but Greek was no longer understood. Society was insecure, but it nonetheless produced magnificent barbaric jewellery, Romanesque cathedrals, medieval illuminated manuscripts,

Beowulf and Norse sagas, and Viking ships. There were other technical advances in these years, including much greater use of iron. From Asia came horse-collars and stirrups, leading to more efficient use of horses, both in warfare and in place of much slower oxen for hauling. Agriculture in north-western Europe was transformed as heavy soils were ploughed by powerful horse teams; crop rotations that included oats and beans to fuel these teams were introduced, with the effect of improving human diets too. Without stirrups, cavalrymen had been in danger of falling off if they did more than harry the enemy; with them, they were secure enough to be able to charge at each other with lances, and knights in armour became an essential feature of armies. Heavy cavalry was an ancestor of tanks.

The Vikings fearlessly opened the sea roads round the north of Scotland, island-hopping across the North Atlantic: settling Iceland, establishing themselves in Greenland and landing across the ocean in Vinland. As life settled down, water wheels and windmills began to provide power, crucially so in the Netherlands, where reclaimed land could now be pumped dry; churches, cathedrals and formidable castles were built, showing bold innovations in stone vaulting and woodwork; the convenient codex (the forerunner of the modern book) came to replace the scroll, and monasteries developed scriptoria where texts were copied. Carvings in churches and miniatures in the margins of manuscript psalters and devotional Books of Hours show a delight in plants, animals and everyday life – as well as in monsters and the grotesque.

European Christians, now literally full of beans, fought each other constantly, but in 1095 Pope Urban II launched them on a series of crusades, predominately against Islam but also against Jews and heretics nearer home, such as Cathars. At the western end of the Mediterranean, Muslim territories in Spain and Portugal were gradually reconquered from now disunited Moors, in the process losing the cosmopolitan character that had made the cities there centres of translation and intellectual exchange – Muslim Córdoba had been the greatest city in Europe. Most Muslims and Jews who refused to convert to Christianity were expelled after Ferdinand of Aragon and Isabella of Castile united Spain in 1492; the process was completed under Phillip II in the 1560s. 'Conversos' who remained were distrusted and liable to investigation by the Inquisition. Perhaps there, and certainly in the East in Byzantium and Islam, the crusades were like the barbarian assaults upon the Western Roman Empire. As part of a violent campaign against internally divided Muslim emirates, Jerusalem was captured in 1099 and for a century became the centre of a Christian kingdom. At first, the Orthodox Christians of Constantinople were viewed as allies by the Western invaders, but they made easier prey than the Muslims, and in the infamous Fourth Crusade of 1204 the city was sacked by its quondam friends, diverted there by Muslim and Venetian diplomacy, and its treasures dispersed, including the famous bronze horses to Venice. Although the 'Franks' were in due course driven out, the Byzantines never fully recovered from the experience; as a consequence, and despite losing ground in the West, Islam gained it in the East. Saladin, artfully uniting the Muslims, and his successors expelled the crusaders from their mainland kingdoms, and in 1453 Constantinople fell to the Ottoman Turks, who then conquered the Balkans and Hungary and would eventually besiege Vienna. As in other prolonged wars, there were long intervals of truce, trading and even alliances, when the Westerners developed more civilised habits, better medical and legal practices, and greater curiosity about the natural world thanks to the influence of their neighbours (most learning went in that direction), until zealous reinforcements arrived for one or both sides and fighting began again.

Muslims had by then long been cultivating science. Caliphs in Baghdad in the eighth and ninth centuries had patronised learning, and inquisitive Muslims were drawn to Greek natural philosophy. Their intermediaries were usually tolerated Nestorian Christians, 'heretics' happier under Islam than under Byzantium, and the texts they translated might go through another language like Syriac on their way into Arabic. Armed with Greek learning, Persian and Arab Muslims carried forward logic, mathematics (notably algebra), alchemy, astronomy, optics and medicine: among their names, corrupted by subsequent Western translators, are the mathematician Omar Khayyam (c.1048–c.1122), the astronomer al Biruni (973–1048), the physicians Rhazes (865–925) and Avicenna (980–1037), the alchemist Jabir (c.721–c.815), and the philosophers al Kindi (c.800–c.870) and Averroës (1126–98). Their interests were encyclopaedic, and their work went way beyond commentary upon the translated texts; observatories and hospitals were built around the Islamic world, and fresh phenomena were carefully recorded there.

Along with algebra, Muslims took up and developed from the Hindus the 'Arabic' number system that we use today. They invented the art of distillation and perfected the astrolabe, an instrument for taking celestial altitudes and for showing where the Sun and planets will be on any particular day. Nevertheless, by the end of the twelfth century, this Arabic Renaissance had lost momentum and philosophers were regarded with suspicion by the orthodox: Averroës died in exile in Morocco, and by 1258, when Baghdad fell to the Mongol invaders, the great days were over, and the lamp of learning was passed to the Franks.

Meanwhile, in the West, Isidore of Seville (560–636) had assembled an encyclopaedia, *Etymologiae*, which became a standard work for Western scholars along with the *Natural History* of Pliny the Elder (23–79); and at Jarrow in north-east England, Bede (c.673–735) not only wrote the *History of the English Church and People* for which he is still remembered today, but also mathematical works, the most celebrated being the *Computus*. At Easter, the Church celebrates Jesus' resurrection, but the date for this festival, the most important in the Christian calendar, varies because he was crucified at the Jewish Passover, the date of which is determined by the first full moon after the vernal equinox. Calculating the date involved the intersection of the (strictly incompatible) Jewish lunar calendar and the Roman solar one, named Julian after Julius Caesar. It was far from straightforward: the Irish, who had evangelised among the pagans in Scotland and northern England, computed the date differently from the authorities in Rome: they were using an older, eighty-four-year cycle instead of the nineteen-year one. Other Churches differed, too. Bede's calculations marked the resolution of this major source of controversy within the Western Church. Astronomy and Christianity came together in the determination of the vernal equinox, the motions of the Moon and even the date of the Creation. Manuscripts of Bede's calculations proliferated, sometimes beautifully decorated, and remained important – especially, but not only, in England – for five hundred years. But compared with what was happening in Islam, all such activity was provincial and unsophisticated.

In the twelfth century, a renaissance based upon Arabic knowledge duly began in the West with the translation into Latin, in Sicily and Spain, of Greek texts that had long been unknown in the West, along with Arabic writings. Wandering scholars such as Adelard of Bath and Gerard of Cremona produced Latin translations of originally Greek texts via the very different language of Arabic. On this linguistic journey, accuracy prevailed over idiomatic fluency: the Latin was clumsy. After initially finding them-selves overwhelmed by this influx of knowledge, Europeans gradually assimilated it (not without difficulties) in cathedral and monastic schools, and then in newly invented educational institutions known as universities. First in Italy, at Salerno and Bologna, then in Paris and Oxford and other cities, the latter developed around groups or 'colleges' of scholars who came together to master texts and to debate, and then to petition for charters from popes and kings to ensure their independence, legal status and survival. In their philosophy faculties they taught – through lectures, commentaries and formal disputations or debates – the classic seven topics that went back to late Antiquity, transmitted through Boethius' *De Consolatione Philosophiae* (c.480–524): the 'trivium' of grammar, rhetoric and logic (whence our word trivial), and the 'quadrivium' of astronomy, geometry, arithmetic and music (the most mathematical of the fine arts). Then, in the higher faculties, men imbibed the theology, law or medicine that would prepare them for the learned professions. Attempts to forbid the teaching of Aristotle as incompatible with Christianity foundered in the light of the new knowledge that study of these 'heathen' works brought.

Two great Dominicans, Albertus Magnus (c.1200–80), now officially the patron saint of scientists, and Thomas Aquinas (1225–74), synthesised Aristotelian philosophy and Christian doctrine, and their views became the orthodoxy against which Galileo and the other heroes of the Scientific Revolution would have to strive. They propounded the doctrine of 'accom-modation', meaning not that the Bible must be accommodated to modern thought, but that instances where the Bible was apparently contradicted by science showed that God's message had been adapted to those to whom it was addressed. The Bible was not a scientific textbook, but (as a witty cardinal later put it to Galileo) was there to tell us how to go to Heaven, not how the heavens go. Ultimately truth would nevertheless come from Scripture and its priestly interpreters, and science was really a system of hypotheses judged by how well they worked. With hindsight, Aristotle's physics, Galen's (c. 130–c. 201) anatomy and Ptolemy's (c.90–c.168) astronomy may seem defective, but in the thirteenth century they potently reinforced the ideas that nature was law-abiding, that the workings of the human body could be understood and that the Earth was a sphere. Moreover, the Earth was a mere point in a vast cosmos, an enormous mechanism in which future events like eclipses could be predicted by those who understood it.

As the Greek language came to be known again in the West, direct translations could be made; and the fall of Constantinople brought more manuscripts into circulation. Ptolemy had been renouned for his astrology, a widely respected science; now his geography (of which more later) and astronomy, with the planets circling the Sun in epicycles like a giant fairground waltzer, or the wheels within wheels of Ezekiel, had to be mastered. The same process happened in relation to Plato, whose stunning dialogues now joined the *Timaeus*, the only one of his works previously familiar in the West. There were exciting books on 'natural magic', said to be by the Egyptian god Hermes Trismegistus, and mathematical works by Archimedes (?c.287–212 BC). In what we call 'the Renaissance', from the fourteenth century, the Platonic emphasis upon the humanities, Archimedean delight in mathematics and Hermetic quests for power over nature played an important part. Humanists sought to write elegant Greek and Latin verse and Ciceronian prose, derided the clumsy and stilted style of earlier translators and, following the mathematician Peter Ramus (1515–72), who valued reason above authority, scorned the logic-chopping of the Dominican scholastics (or 'schoolmen') and, indeed, the down-to-earth philosophy of Aristotle. They made a classical 'liberal' education the norm for gentlemen until well after 1900. At first this new learning flourished in courts and the academies they promoted, and in circles associated with printer-publishers and booksellers, rather than in universities, where the syllabuses remained conservative. Platonic beliefs that mathematical harmonies underlie natural processes, and that ordinary things are imperfect realisations of ideal forms, were very attractive. The next task was to bring them down to earth, to combine them fruitfully with an empirical, experimental style, carried on in a workshop or quarry, or on board a ship, where practical experience and technique, the craftsmen's hand and eye and rule of thumb, were paramount and respected. Something new and exciting was beginning.

The gap between the old and the new ways of looking at the world is best followed thematically: we begin with mapping, where it is vividly illustrated. Medieval world maps, like the one in Hereford Cathedral, were circular and had east at the top, where the Sun rises. Round the edge was the ocean; the Mediterranean went downwards from the middle to the bottom of the map, where, at the Pillars of Hercules, the Sun set and the world ended. Asia occupied the top half of the map; Africa the bottom right sector, and Europe the bottom left. The effect is a bit like a 'T' inscribed in a circle: as such, they are called 'T-O' maps. The British Isles are in the ocean to the north-east; but the Hereford map was never intended to help the pilgrim get from the English Midlands to Rome – for that, there were itineraries, like that of the historian Matthew Paris (1200–59), now in the British Library. Maps were made to display the history and geography of the created world, based upon the Bible, and were there to help us get to Heaven. In the middle is Jerusalem, where the central events of world history took place: the Crucifixion and Resurrection of Jesus. Ever since, the world has been waiting for his Second Coming. Surmounting the circle is Christ in Majesty, coming to judge the world. Within the map, the Garden of Eden is at the top, with Adam and Eve eating the forbidden fruit. Nearby in Paradise are Enoch and Elijah, who did not die but will reappear at the end of time to be slain by Antichrist and then resurrected with the righteous. As we go down through Asia and the centuries, we see the events of Jewish history with some embellishments: thus, to our left, we see the man-eating giants Gog and Magog, imprisoned by Alexander the Great on a walled-off peninsula, who will in the Last Days be released to work woe upon mankind.

Around the map, where today we might expect to find the four letters N-E-S-W, standing for the cardinal points of the compass, the word MORS (Latin for 'death') appears; and death, the consequence of Adam and Eve's sin, meant (for good church people) release from this world of toil and grief. The course of empire moves westwards through this history, from sunrise to sunset, Eden to Babylon to Rome and onwards: history is linear, from alpha to omega, and is the fulfilment of God's inscrutable plan. There is not, however, so much in the bottom half of the map, which represents the end time, our time, and the waiting and watching for the Apocalypse and Jesus' return, where the Sun sets over the Pillars of Hercules at the very bottom. For us, the term 'apocalypse' represents a disaster of inconceivable proportions – but for our ancestors it represented the Revelation of God at the end of time, something to look forward to, when justice would be done, every tear would be dried, the righteous would enter into everlasting bliss and the wicked be cast down with Satan, bound at last, into the pit. Jerusalem is often shown not as the cartographer imagined it to have been in Jesus' time, but as the New Jerusalem come down from Heaven, with golden walls, where Jesus will reign.

On the wall inside a cathedral, or in a psalter or devotional book, these circular maps focused meditation on the transitory character of material things, and on space and time, which had only come into being when God made the world, and would cease with it. The renewed Heaven and Earth would be eternal and timeless. The secular and the sacred were not, for these mapmakers, separate spheres; and the Bible was not 'a religious or sacred book', one of a class including the Ramayana, the Koran and the Granth, but a unique source of truth when properly interpreted. Such maps went on being made down to the middle of the fifteenth century, not long before our story begins. No wonder the rediscovered Aristotelian world – godless, eternal and deterministic – caused alarm among many: something serious was in danger of being lost.

The idea that maps should be primarily an aid to seeing where places are and should provide assistance in getting from one place to another was slow to catch on. Indeed, like our ancestors, we still expect our maps to indicate sites of historic significance; the shape and size of continents may be adjusted to reflect populations or economic importance rather than strict topography. London Underground maps are useless if you want to find your way about the city at ground level. Sailors had simple, practical requirements, and in the Mediterranean developed charts to help them navigate. As magnetic compasses came into use from the twelfth century, these charts were crisscrossed with rhumb lines for direction-finding that radiate from the points of 'wind roses', showing all the points of the compass. Using a ruler and dividers, a pilot could plot his course on such a chart. By about 1450 chartmakers knew that needles did not point to true north and took account of that fact; and charts, called in northern Europe rutters or later 'waggoners' (from the Dutch cartographer Lucas Waghenaer, 1534–1606), came into use on the Atlantic. Terrestrial magnetism remained a mysterious phenomenon yet to be properly investigated; meanwhile, on land, estate and city maps were coming into use, incorporating surveys and bird's-eye views. To map a whole country was a different proposition, undertaken from the sixteenth century onwards but demanding increasingly accurate instruments and teams of observers.

A new kind of world map was, however, available as the Scientific Revolution began in the later fifteenth century. In order to map a spherical Earth on a flat surface, Ptolemy had proposed a conical projection with straight longitude and curved latitude lines; no examples survive from Antiquity, but the first printed version was published in Bologna in 1477. His data were best for the Mediterranean region, and there are serious distortions as we get further from it: Sri Lanka is very large and India rather small, Scotland is bent round to the east, and so on. But as a bold attempt to plot ancient geographical knowledge it was astonishing, and the printed version soon provoked attempts to do better as new discoveries came in. Ptolemy's Africa included the Mountains of the Moon where the Nile rose, tantalising geographers and explorers for centuries. The spate of new maps – a major feature of the Scientific Revolution – fulfilled the needs of sailors and travellers (actual or armchair) as ancient exemplars, geometry, dead reckonings on land and water, latitudes and longitudes newly confirmed by astronomical observations, and the new technology of printing were all brought together in the printing house.

The Renaissance meant a huge admiration for ancient achievements, particularly in Italy because Greece was inaccessible for most Westerners. But once the ancients' ways of writing, building, painting and sculpture were understood, it became possible to break away from, and seek to surpass, classical models. Maps are one example: Ptolemy's would no longer do, when the 'torrid zone' around the equator was proved to be habitable after all, a new world was found between western Europe and eastern Asia, and humans were encountered in the southern hemisphere, the Antipodes where everything was supposed to be topsy-turvy. Oceanic voyages were not a European innovation or monopoly: Polynesians had settled islands dotted all over the Pacific, culminating in their colonisation of what we call

New Zealand. Malaysians had navigated the Indian Ocean, and had settled the enormous and previously uninhabited island of Madagascar, with its extraordinary flora and fauna. Later, Arabs had sailed their dhows there and to East Africa, and across the Persian Gulf to India; the stories of Sinbad the Sailor in the *Arabian Nights* are based upon their exploits. Junks exploring from China had also reached East Africa. Overland, along the Silk Road from China to the Ottoman Empire and on to Venice, had come kites and the recipe for pasta that transformed Italian cooking, as well as luxury goods and reports of devices such as canal locks, and pumps.

Trade was opening up the world, and the Hellenistic tradition of mapping was not the only one, for the Malaysians, the Chinese, the Arabs and the Polynesians had their own ways of plotting journeys and surveying lands. Incorporating their stories, connecting the realms of the created order, world history, epic voyages and travellers' tales with exact observations and navigators' needs, and getting them down into map form, was a task for those at the outset of the Scientific Revolution. Knowledge is power, and it would have suited the kings of Portugal and Spain to keep data from their voyages secret so as to exclude Dutch, French and English 'pirates' from their colonial waters; but publishers in the Netherlands won that round of what has been a continual battle to keep science open and public.

When oceanic voyagers made their landfall almost everything was unfamiliar, and they collected avidly from the animal, vegetable and mineral realms, bringing back samples as well as native inhabitants, who were either persuaded to come along with them or simply kidnapped. The Portuguese, having made their way steadily down Africa, found at the Cape of Good Hope a climate not unlike their own, but completely different animals and a very rich new flora. They encountered peoples living very differently, speaking very distinct languages and from different ethnic groups. Sailing on to the north, they named Natal on Christmas Day, and met a more familiar world as they encountered Arabs in East Africa. Crossing the Indian Ocean to Calicut, they found themselves in a country familiar by repute, and promising commerce. Vasco da Gama could not establish a foothold or 'factory' in what turned out to be hostile territory, and returned to Lisbon, whence in 1500 King Manuel dispatched Pedro Alvarez Cabral (who landed in Brazil and claimed it, on the way), in 1502 da Gama again, and then Afonso de Albuquerque. With small bodies of soldiers, they defeated local rulers and set up bases in Mombasa and, in 1510, Goa. Thence they expanded further east, but in 1578 King Sebastião decided to attack Muslims nearer home, invaded Morocco and was killed along with most of his army at Alcazar-Qivir. Two years later Philip II of Spain took over the country: for sixty years Portugal lost its independence, and its factories and colonies became fair game for the Dutch, fighting at home for their independence, and the English following their defeat of the Spanish Armada in 1588.

Vasco da Gama's voyage was commemorated by Luis Vaz de Camões (1524–80) in his poem *The Lusiads*, an epic in which the hero and his crew bring Christianity to the East with the assistance of Venus and despite the opposition of Bacchus and Neptune, as well as Muslims whose trade routes were threatened by these interlopers:

> Though Fortune stands upon a tott'ring ball . . .
> Renowned people you shall never lack,
> Wealth, valour, fame, till the worlds henges crack.[1]

It did not happen in quite that way. The voyages and campaigns were desperate adventures for a small country, involving heavy loss of life; accounts of shipwrecks and the exploits and misfortunes of the castaways were later collected and published as *The Tragic History of the Sea* (1735–6). Lessons might be learned. For example:

> The reason for the loss of this great ship, which is the same as that for nearly all of them. . . . What route should be followed and what avoided, what preparations should be made in view of its length and difficulty, how to treat and deal with Kaffirs . . . and their barbarous nature and customs.[2]

As well as the hazards of storms, leaks and uncharted reefs, after 1580 the Portuguese had to face attacks at sea and on land by the Dutch and English, eager to capture their ships, and to establish bases of their own, taking over local as well as intercontinental trade. The Dutch with their shallow, broad-bottomed 'flutes' had come to dominate the Baltic trade, and their commercial practice, banking, exchange and bookkeeping, maintained in the East, promoted objectivity in classification. In these various tales of triumphs and catastrophes there is both demand for accurate mapping and charting, dependent on astronomical observation and calculation, and much material for natural history and the study of mankind. All this came out of commercial enterprise under royal or republican aegis in the form of trade with India and Africa. We should not allow ourselves to forget that the 'barbarous Kaffirs' of Natal, who helped or plundered the shipwrecked, in due course became themselves a lucrative and vile part of this trade, bought and sold as slaves.

Columbus, spurred by his ill-founded hypothesis that the Earth was small, sailed the other way, to the West Indies, thinking he must have reached the fringe of Asia. It was hard to believe that there was an unknown, even unsuspected, continent: once that was accepted, it made transatlantic voyages of discovery even more exciting. The position, limits, history, geography and natural history of this vast territory were all unknown. Aided by disease to which the natives had no resistance, and by local dissensions, small bodies of Spanish desperadoes conquered Mexico and Peru. From the start, religious orders tried to mitigate the violence of conquest and convert the people they met to Christianity; and missions, bishoprics and universities were set up in what became colonies. As well as treasure, notably a steady stream of gold and silver, much information about the resources of the New World came to Seville, and although these might seem state secrets to be kept from prying foreign eyes, word got out. Potatoes, corn (maize) and chilli peppers entered menus in Europe and beyond. The smoking of tobacco, loved and hated from the start, gave Europeans a new habit. There were some geographical mistakes: 'turkeys' were thought to be Turkish, rather than from North America, where after an abortive attempt to found a colony in 'Virginia' (Roanoke) the British succeeded in establishing themselves further north in modern-day Virginia and Massachusetts, while the French settled in Canada, where there were furs to trade as well as cod. The pope divided up the world between the Portuguese and Spaniards with a line (to be based upon astronomical observations) down South America and Asia, but other nations were unimpressed. Factories and bases, settlements and colonies, under kings or chartered companies, grew into seaborne empires with governors and vice-roys. Their control, exploitation and defence became a crucial feature of European statecraft from the sixteenth century down to our own times.

Science became global when, in 1519, Ferdinand Magellan (c.1480–1521), a Portuguese in Spanish service, set out with a fleet of five ships to reach the Moluccas from the west. He navigated the strait that now bears his name in Patagonia and sailed for ninety-eight days across the Pacific. Beset by scurvy, the fleet was by then down to two ships: one was captured by the Portuguese as having strayed into their part of the world, but Magellan's, the *Victoria*, reached the Philippines (where he was killed in a skirmish) and made it back to Spain in 1522. Among the survivors was Antonio Pigafetta (c. 1491–c. 1535) from Vicenza, a gentleman volunteer with a lively curiosity about peoples and their languages, who kept a record of the voyage, thus inaugurating a tradition that was to culminate in Joseph Banks (1744–1820) and Charles Darwin (1809–82). He described flying fish ('very good to eat') and how bonitos chase them, 'a marvellous and merry thing to see'; and from a 'Patagonian giant' he collected a vocabulary, much of it concerning body parts but including 'The Great Devil, Setebos', who later featured in Shakespeare's *Tempest*.[3] For three hundred years these giants of Patagonia excited our ancestors as the subject of an evidence-based traveller's tale, but the sober truth seems to be that many of them were 6 feet (182 cm) or more tall at a time when Europeans were generally shorter.[4] Globes were covered with printed gores, tapering strips of paper that converged at the poles. By 1526 Magellan's track was shown on gores printed, probably in Nuremberg, to form a globe that features in Hans Holbein's (c.1497–1543) famous painting *The Ambassadors* (1536). All these voyages aroused wonder at the astonishing variety of things, and at previous ignorance; and proved that European technology in the form of ships and firearms could dominate the world, bringing power, riches and the benefits of trade. They brought new confidence, and provided a metaphor for scientific discovery itself.

Not only giants but extraordinary creatures were fascinating. The rhinoc-eros, unseen in Europe since Roman times, had been known by repute for its ferocity, especially towards elephants: the Parisian philosopher Peter

Abelard (1079–1142) was called the 'Indomitable Rhinoceros' for his defence of Nominalism. In 1514 Afonso de Albuquerque sent one to Lisbon from India as a gift for the royal menagerie. In the following year Albrecht Dürer (1471–1528) published an engraving of it, based on a report by others, and his magnificent beast with splendid Renaissance armour and an extra horn between its shoulders drove out more accurate representations for almost three centuries. Science begins with facts, but method was required to verify and classify them (filtering out giants and extravagantly armoured rhinoceroses), find explanations and then be in a position to use them like travellers with a map. In the face of new knowledge, Ptolemy's maps were no longer adequate; meanwhile, astronomers had been concerned about awkward features of Ptolemy's wheels within wheels, and about the poor fit of his model with the much less accurate Aristotelian nest of spheres, displayed on the clock at Wells Cathedral in England and generally accepted by non-astronomers as explaining heavenly motions.

We find in Renaissance painting something of the empirical spirit that was transforming mapping. Icon painters in the East, and those depicting religious scenes in the West, had not sought to be realistic: after all, they were painting scenes from the Bible or from legends connected with it, saints preaching or performing miracles, grisly martyrdoms, or demons in Hell and angels in Heaven. They were not trying to paint portraits or scenery, but to inspire interpretation (historical, allegorical or personal), awe and reflection, penitence and amendment of life: backgrounds might be plain or gold, saints were shown with haloes and their characteristic symbol, and, as in Assyrian sculptures, the important characters were larger than minor ones. In the early Renaissance, artists in Flanders and Italy began painting backgrounds that were identifiable places. Then, in 1425, Filippo Brunelleschi (1377–1446) started to use one-point linear perspective and, in about 1435, Leon Battista Alberti (1404–72) wrote of the plane of the picture being like a window through which the illusion of three dimensions is obtained. Artists took to using devices such as the mirror and the camera obscura to assist them in achieving accurate representations and trompe-l'oeil effects. Sometimes the subject of a portrait, or the donor included in a religious picture, holds or wears spectacles, for skilled workers in a new trade began to grind lenses for reading glasses that allowed elderly intellectuals to go on working. Lenses focusing diverging beams of light would also come to interest artists. Patrons began to commission portraits (though we should not imagine kings giving long sittings for painters) and secular scenes as well as religious pictures, and artists sought to draw viewers into the picture, as though they were meeting the person portrayed, or were part of the scene displayed. An experimental spirit and new realism developed alongside allegorical and symbolic representation: scenes from the Bible were brought into the here and now. Artists moved up in the world; no longer just artisans but would-be gentlemen, desirable at court, they began to sign their pictures.

On T-O maps, the ocean surrounds the continents and the Mediterranean is prominent; but land, not water, fills the map. On the new world maps of the sixteenth century, water predominates, even though the enormous and hypothetical southern continent, Terra Australis Incognita, is there to balance the land in the northern hemisphere. Water, a necessity for life, can be and has been thought of in many ways: notably as a symbol, as an element and as a substance. As a symbol, it is ambiguous: we need it, but it may drown us. In the Bible, water cleans and cools, but it is also chaotic and threatening. Baptism is the washing-away of sin, but it is also a little death as the candidate is plunged beneath the surface, to emerge and begin a new life. Though the Children of Israel had passed dry-shod through the Red Sea, in John Bunyan's (1628–88) *Pilgrim's Progress* (1678) we are reminded that we must pass through the Jordan, the waters of death, before the trumpets sound on the other side. In the Bible, Noah and Jonah were preserved by the Ark and the whale, and believers hoped the Church would save them. In the New Heaven and Earth of Revelation, there would be no more sea. In classical myth, too, Hades had its grim rivers, Acheron, Cocytus, Lethe, Phlegethon and Styx; while Narcissus drowned in the clear pool that reflected him. Until really clear glass became available, water was the primary substance that reflected and refracted light.

Water was, with Earth, Air and Fire, one of the ancient elements; cold and wet, it was present in everything terrestrial, and dominant in liquids and transparent bodies. It might even be the source of everything, transmuted, for example, by living organisms into other elements. This idea provided a framework for understanding the weather, minerals, and our bodies and temperaments. Water is a great agent of change, wearing away rocks; rivers, making irrigation and intensive agriculture possible, created the sites of the first cities; and water is either a barrier or a highway for travellers, depending upon whether they have ships and boats. The river that endures though the water flows on has provided philosophers since Heraclitus (fl. c.500 BC) with an example of universal flux; while the rainbow, a sign sent to Noah and his descendants that God would not again flood the whole Earth, was a puzzle for Muslim and Christian thinkers who sought to account for it. The work of Alhazen (c.965–c.1040) on this and other optical problems was taken up by the Silesian friar Witelo (exact dates unknown), writing on perspective in the 1270s; and his younger contemporary, the Dominican Theodoric of Freiburg (c.1250–c.1310), experimented using a spherical glass vessel full of water to simulate a raindrop. Such flasks were used by doctors who, in making a diagnosis, would examine their patients' urine by holding it up against the light. Theodoric put in clean water, and with the light behind him raised and lowered the flask and got flashes of colour. The rainbow was thus generated by reflection and refraction when the Sun shone on raindrops, the colours somehow forming arcs. This was an explanation in principle rather than in detail, leaving more work to be done; but it raised a problem for expositors – how had there been no rainbows before Noah? In the seventeenth century Thomas Burnet (c.1635–1715) offered the suggestion that there had been no serious rain in those days, but only mist. The poet John Keats (1795–1821) blamed Newton for making rainbows ordinary; but in fact this happened much earlier.

More prosaically, water is a substance with curious characteristics – notably the way it expands on freezing. The chemists in the eighteenth century who discovered its composition investigated its anomalous properties, and water would subsequently be the key to understanding much about chemical affinity. This story of symbol, element and substance might seem to accord with the idea of Auguste Comte (1798–1857) that understanding (in both individuals and societies) proceeds through three stages: theological, metaphysical and positive – the last being grown-up and scientific, and so superseding the others. We do not need to believe that: in the seventeenth century there were metaphysical poets like John Donne (1572–1631) playing with resonant words and ideas, but theology, the

creative arts and science were not separate spheres, and natural philosophers did not intend to turn the world grey with their breath. Symbols and
signs retained their power, and science was indeed to add to them; but just
as the Protestant Reformers disposed of symbolism, icons and allegorical
readings of the Bible in favour of literalism, so the new men of science
sought sober plain words and empirically testable hypotheses. They were
duly then followed in the late seventeenth century by preachers in a new
and plainer style of rhetoric.

People had dropped, pushed and rolled things, thrown stones, balls and
javelins, and shot arrows for millennia, unworried by how motion, especially of projectiles, was possible. Aristotle knew from the experience we all
share that, unless pushed or pulled, moving objects come to a halt. He also
knew that the heavier things are, the harder they fall. Some motion was
natural, going to the proper place: downwards for earthy or watery objects
and upwards for fire. The changeless heavens (made of quintessence)
moved in systems of perfect circles, propelled by the Prime Mover; the
Earth was a sphere at the centre of the universe, where all the heavy base
matter had collected, and was subject to vicissitudes of all kinds. Ancient
Greeks knew that falling bodies go faster and faster, and Aristotle thought
that their speed was proportional to their weight. This view was contested
by John Philoponus (490–570), who seems to have dropped different
weights from a tower in Alexandria and found they reached the ground at
the same time: an experiment later attributed to Galileo using the Leaning
Tower of Pisa. Philoponus introduced the idea of 'impetus' imparted to
bodies as they were taken up a tower: when they were released, as the
impetus was used up they fell faster and faster towards their natural place.
Very critical of Aristotle, and denounced as a heretic for his theological
ideas, Philoponus was little known later, and the exact behaviour of falling
bodies remained uncertain.

Falling was natural, but unnatural motions on the Earth's surface
required force: for horses and carts this was obvious, but the flight of
projectiles was particularly tricky to explain. One possible account involved
the air, cut away in front of the projectile and closing up behind it. Jean
Buridan (c.1300–58) and Nicole Oresme (c.1320–82) in Paris found this
unsatisfactory. They took up and developed the idea of impetus, supposing

that it was imparted to the projectile by the wrist, the bowstring or the gunpowder, and was gradually used up in its flight. When the impetus was exhausted, the stone, arrow or bullet fell to the ground. This was an important step towards the mechanics of Galileo and Newton; but again it was an explanation in principle rather than something firm and detailed, and there was no clear idea of the trajectory of projectiles.

Meanwhile, Robert Grosseteste (c.1168–1253) had as chancellor at Oxford lectured to the newly arrived Franciscans, and induced them to take up mathematics and natural philosophy. His particular interest was optics, for he saw light as both religious symbol and active agent; as bishop of the enormous diocese of Lincoln he was a very powerful man, and he passed on a commitment to experiment to Roger Bacon (c.1214–92). Subsequent mathematicians at Oxford found the way to handle what they called uniformly difform motion, our uniform acceleration. Thomas Bradwardine (c.1290–1349), later archbishop of Canterbury, and his collaborators at Merton College, William Heytesbury (c.1313–72/3), Richard Swineshead (fl. 1340s–'50s) and John Dumbleton (d. c.1349), distinguished kinematics, the study of motions, from dynamics, the science of forces. Their 'mean speed theorem' showed that the distance covered by an accelerating body was the same as if it had moved at the mean speed for the same time. This could be seen to generate the law of falling bodies that we now have, and that is credited to Galileo. The capacity to idealise, neglecting particular and adventitious circumstances and going for the simple mathematical relationship, did not leap over the centuries from Archimedes to Galileo: it was there and practised before 1400 by these various thinkers, generating ideas and data for a new and powerful synthesis: the mechanics of the seventeenth century.

Before printing, books were luxuries and libraries were small. Monks used to copy manuscripts, illuminating them with handsome initials and perhaps also illustrations in the text or the margins – such manuscripts are among our great treasures. Many, though, were more workaday, and from 1400 paper, invented in China and introduced to Europe through Islam, superseded for most purposes expensive animal-skin vellum or parchment. By 1450 lay scribes in commercial scriptoria were producing texts for students as well as prayer books, poetry and even romances. On each occasion that a manuscript was copied, some errors (or emendations) would inevitably be made; the genealogy of manuscripts can be worked out that way. In search of the oldest, and thus most authentic, source, a scholar would have to travel; and, indeed, though Bede never left Northumbria, wide reading often required journeys, and thus time. Printing, and the marketing of books through international fairs, changed that, making available some texts that were exceedingly rare, such as Lucretius' (c. 99–55 BC) *On the Nature of Things* (*De Rerum Natura*), of which only one manuscript had survived. The cost of books fell with mass production and marketing, and the sciences were (along with Protestantism) major beneficiaries. Indeed, we can say that we owe the success of the Scientific Revolution and the Reformation to printers.

The roles of printer, publisher, stationer and bookseller were not separated, and the production of books involved mechanical skills, financial know-how and good contacts. Printing houses became intellectual centres; but most of the books published were old favourites, for which there was a known demand. Thus, in astronomy, about 1230 Sacrobosco (John of Holywood, c.1195–1256) wrote *The Sphere*, concerned with the motions of the planets. This became a standard text: numerous editions were printed before 1500, though by then it was far from being up to date. Sacrobosco was important, however, for commenting on the errors of the Julian calendar, which since Bede's day had got about ten days out: the equinox no longer fell where it should because every fourth year was made a leap year. He had proposed omitting a leap-day every 288 years to get things right. Various church councils considered how to avoid a reignited Easter controversy, until in 1475 Pope Sixtus IV invited Regiomontanus (Johann Müller, 1436–76) from Königsberg to Rome to assist with the reform. Regiomontanus had mastered Ptolemy's *Almagest*, published astronomical tables and established in Nuremberg a workshop making astronomical instruments. He died young, soon after getting to Rome, but the project went on in its stately way, culminating in the calendar drawn up by the Jesuit astronomer Christopher Clavius (1538–1612) and promulgated by Pope Gregory XIII in 1582. That year, 4 October was followed by the 15th and henceforward centennial years would not be leap years unless, like 2000, they were divisible by 400. Years started on 1 January instead of 25 March. Gradually, this system prevailed, first in Roman Catholic countries, then in the Netherlands, in Britain in 1752 and in Russia only in 1918 (when Lenin began the October Revolution, the rest of the world had progressed into November). The existence of two systems made for confusion about dates as some states used the old and some the new style. The first months of the year were often shown as, for example, 1625/6: legally it was 1625 in England, and 1626 in France, and there was also an eleven-day difference, so a letter dated 9 January 1625 from England could be a reply to one of 10 January 1626 from France. Meanwhile, the coming of mechanical clocks had meant general adoption of the Roman idea that the day began at midnight: a difficult time to spot, and thus less natural than dusk or dawn – or noon, which remained the beginning of the day on board ship until the end of the eighteenth century.

Pursuing the calendar has taken us way beyond the Renaissance, but the story involves young Copernicus, who went to Italy to complete his education in canon law, in 1496–1500 and again in 1501–3, at Bologna, Padua and Ferrara. His interest in astronomy was kindled there. Back in remote Frauenburg (now Frombork), where he was a lay canon and administrator in the cathedral and his uncle was bishop, he could develop that interest because printed texts were available: it would not have been possible a hundred years earlier. As well as diagrams and tables, some books such as the compendious *Margarita Philosophica* of Gregor Reisch (c.1467–1525) had volvelles, paper circles that rotated over each other, making a kind of astrolabe for calculations. Thanks to his books, notably Regiomontanus' works, Copernicus mastered Ptolemy's Earth-centred system and had the confidence to be dissatisfied with it: the Moon should have varied considerably in size as it went round in Ptolemy's epicycles, and for the planets Ptolemy had used a dodge, the equant, that made irregular motion seem uniform. Copernicus took the system to bits and put it back together again, more elegantly. From ancient Pythagoreans and Aristarchos of Samos (310–230 BC) he adopted the Greek minority view that the Sun was the centre of the world, an idea that Cardinal Nicholas of Cusa (1401–64) had also taken up, writing on 'learned ignorance' and suggesting that the cosmos was a sphere with centre everywhere and circumference nowhere. We shall return later to Copernicus' astronomy, the point here being that it was now possible to do serious science away from great centres, aware through printed books of what was going on, and able to promulgate a new theory by publishing a book. In the subsequent history of science, such relatively isolated people have been responsible for some major developments.

Clavius was a Jesuit, Reisch a Carthusian, Cusa a cardinal and Copernicus a lay canon: but science was not only the preserve of church-based intellectuals. Renaissance Europe was abuzz with commerce and industry; the republics of Venice and Florence were ruled by merchants and bankers, anxious for useful knowledge, who also held power in great northern cities like Antwerp, Amsterdam and London. Craftsmen, learning by apprenticeship, had accumulated many skills and much tacit knowledge of materials. It is a feature of the history (and present state) of science and technology that inventions frequently precede the scientific knowledge underlying them, and that natural philosophers learn much by reverse engineering, investigating devices which they may subsequently be able to improve. Craftsmen's activities were a catalyst for the Scientific Revolution, as some learned to read and write, and as more highly educated contemporaries began to be interested in such activities and to record them in books. This happened for printing, as we might expect, but it was particularly notable in mining and metallurgy.

Vannoccio Birunguccio (1480–c.1539) came from an iron-founding family in Siena, where he worked at the mint and the arsenal. He cast cannon in Venice and then in Florence, and in 1536 went to Rome where, in 1538, he took charge of the papal foundry and became director of munitions. In 1540 his *Pirotechnia* was published, describing how these things were done. More famous was the *De Re Metallica* (1556) of Agricola (Georg Bauer, 1494–1555), translated a century ago by Herbert Hoover, future president of the USA. Agricola was a learned man, a philologist and physician, who held medical appointments in his native Joachimsthal and then in Saxony at Chemnitz. The illustrations of mining machinery and of miners at work in his great book, published posthumously, are magnificent, conveying wonder along with technical information about the great water wheels and pumps that drained the mines. Miners from Germany found their way to England, where by this time the coal trade from Newcastle was becoming important, and brought with them the terms 'fire-damp' and 'choke-damp' for the noxious gases in coal mines. Technical optimism was a feature of the times; as kings sought for gold, their subjects became skilled in the use of iron and other metals, for which they found new uses, replacing wood and leather. Renaissance monarchs like Henry VIII of England (1491–1547) had magnificent armour made for jousting, but smiths also did wonderful work in wrought iron for church doors, for ornamental railings and for weathervanes, as well as making tools and parts of machines.

Agricola's central Europe was in a ferment at this time, with new universities, religious and medical controversy, and alchemy as well as booming industry. He was interested in the possibility of adding metallic remedies to the herbs that were recommended but often ineffective in medical practice. His contemporary Paracelsus (Theophrastus Bombastus von Hohenheim, 1493–1541) studied minerals in the Tyrol as well as medicine at Basel. He described silicosis among miners, one of the first industrial diseases to be identified. Fascinated by alchemy, he hoped that it could be applied medically. He became notorious when he publicly burned the works of ancient physicians as useless, and prescribed metals including mercury, which, though poisonous, turned out to be a specific for syphilis, a new plague supposedly brought home from America by Columbus' sailors. Although Hippocrates (c. 460–377 BC) had written that desperate diseases require desperate remedies, this early chemotherapy did not endear Paracelsus to contemporary doctors, or indeed to the families of those patients who did not recover, and he led a wandering life, widely regarded as a quack. The eminent physician and essayist Sir Thomas Browne (1605–82) wrote scornfully of 'dependence upon the Philosopher's stone, potable gold, or any of those Arcana's whereby Paracelsus that died himself at forty seven, gloried that he could make other men immortal'.[5] Gradually, because they worked and there was nothing else, powerful inorganic compounds made their way into pharmacopoeias. Mystic as well as medic, Paracelsus' independent, innovatory, experimental spirit was important in bringing chemistry into medicine, and for promoting the chemistry of the sixteenth and seventeenth centuries: in place of Aristotle's four elements he proposed 'tria prima' – sulphur, salt and mercury – as the basis for chemistry. In 1830 the chemist and historian Thomas Thomson (1773–1852) saw him as having been responsible for the first chemical revolution.

Learned interest in machinery at this date was not limited to Europeans like Agricola, for in 1637, just before the collapse of the Ming dynasty in the face of the Manchu, a technical encyclopaedia was published in China, *T'ien-kung k'ai-wu* (Exploiting the Works of Nature). It was written by Sung-hsing (1587–1663) and was copiously illustrated with woodcuts. This was precisely the period when Europeans were catching up with Chinese science and technology, and Jesuit missionaries had found a welcome and a niche for Western ideas at court in Beijing. It was to be some time before some Chinese innovations, like the double-acting bellows for a continuous blast and the crankshaft, were to make their way firmly into European devices. But in sixteenth-century Europe there was a fascination with machines, both useful and impracticable (like a rolling mill driven by a smoke jack), and there are superb pictures of them from this time. Water-works, for supplying cities, for transport and for show as fountains were particularly noteworthy.

Carrying goods and passengers in boats and ships was much cheaper than taking them overland, and the building and sailing of vessels for river and canal navigation and for ocean voyages were important activities, involving numerous crafts and skills. Harbours and rivers needed to be dredged, jetties and moles built, and canals dug, with locks (another Chinese invention) between different levels. Across rivers, bridges of ever greater length and boldness were constructed; and fortifications required continual updating as artillery became more and more formidable. The high walls of castles and cities were replaced by squat star shapes that would resist bombardment and allow defenders to mow down attackers with crossfire. The French invasion of Italy in 1494 set off decades of increasingly sophisticated war, as the Habsburgs, the pope and city-states sought to defend and extend their territories. Archimedes was famous for his machines that (for a time) defended Syracuse against the invading Romans; the cathedrals had required of their master masons a mixture of intuitive mathematics, artistry and technical skill; by 1500 all these modern developments brought into notice their successors, the first engineers, including Leonardo da Vinci.

In Renaissance literary circles there was a battle of the books, debate over whether moderns could hope to equal the elegance, wit, pithiness, pathos, flamboyance or eloquence of the best ancients. Universities promoted the writing of Latin and Greek prose and verse, awarding prestigious prizes for the best examples; and Latin, spoken and written, remained the vehicle for learned discourse. Clearly, it would be hard for most of us to rank Renaissance and classical Latin texts; but we can be sure that Milton's Latin verse, of which he was proud, though highly regarded in its time, is not now read. It is harder to be original in a dead language, and originality has been an ever more admired quality in both arts and sciences. Vernacular literatures flourished in the Renaissance, with drama breaking the bonds that Aristotle and other critics had thought should constrain it in the works of Christopher Marlowe (1564–93), William Shakespeare (1564–1616) and their contemporaries – making it absurd to try to rank them against ancient playwrights. The great astronomer's father, Vincenzio Galilei (1533–91), was an eminent musician associated with the beginnings of opera, and in 1607 the *Orpheus* of Claudio Monteverdi (1567–1643) was performed in Mantua. It was an exciting time.

Comparing ancients and moderns in science is different because, without denying greatness to Aristotle, Galen or Ptolemy, one could unambiguously say that by the seventeenth century more was known about motion, about anatomy and physiology, and about geography and astronomy. Moderns used much more iron, and had machines, weapons, foodstuffs and contrivances unknown to the ancients. They had gone through the Pillars of Hercules and sailed, literally and metaphorically, into strange seas. Bacon in *Sylva Sylvarum* (1627), his compendium of curious facts, noted: 'They have in Turkey, a Drink called Coffa, made of a berry of the same Name, as black as Soot and of a Strong Sent.'[6] The moderns soon acquired a liking for coffee, which became important for social and business gatherings. They had other new habits and tastes, too – drinking tea from China as well as smoking tobacco from America. And yet some of the greatest were reluctant to believe that they were innovating rather than recovering lost knowledge familiar to the ancients. Kepler was haunted by the Pythagorean harmony of the spheres, inaudible to mortal ears, which he wrote down in his *Harmonices Mundi* (1619), and Newton believed both that this enigmatic 'harmony' revealed that these sages knew esoterically the inverse square law of gravity, and that interpreting the apocalyptic prophecies of Daniel in the Bible was worth a great deal of his time.

Whereas Shakespeare's plays and Monteverdi's operas are still performed, and the poetry of Dante (1265–1321), Donne, George Herbert (1593–1633) and Milton is still read, nowadays scientists do not study the classics of science. They have become obsolete, their insights (supposedly at least) incorporated into textbooks; for science is a progressive activity, its conclusions provisional and dependent upon available evidence. While there is a huge body of factual knowledge that endures, science is taught through a curious pattern of learning and unlearning that partly recapitulates its history; a good deal of what I was taught half a century ago is now seen as erroneous. And yet the Scientific Revolution was a search for truth and certainty. The latter had been sought in religion, but the Reformation cast all religious doctrine into doubt, and in ancient authorities, but those were quickly seen both to disagree among themselves once really good texts were available, and to be ignorant of the New World across the Atlantic. Opinions, conjectures and prejudices were associated not only with Plato's Cave but also with deferential book learning: natural knowledge acquired in the light of appropriate methods was to be the key to a real understanding, and thus to wisdom and genuine natural philosophy.

The search was on for a new philosophy, a support, a programme and a method for acquiring and ordering natural knowledge so that it could be not only contemplated but also used for what was called the improvement of man's estate. That would be a crucial feature of the Scientific Revolution, distinguishing it from what had gone before; and to that essential underpinning we now turn.

CHAPTER 3

REFINING COMMON SENSE
THE NEW PHILOSOPHY

FROM THE TIME OF the Scientific Revolution, there has been wide expectation that true science would result from correct method, but little agreement about what that might be, and how far it could be learned without all the mass of facts that sciences have to explain. Hoping to make the science of his day accessible, Humphry Davy wrote: 'Science is in fact nothing more than the refinement of common sense making use of facts already known to acquire new facts.'[1] Thomas Huxley (1825–95) proposed that it was 'trained and organised common sense'.[2] When due emphasis is given in either case to 'refinement' and 'trained and organised', these definitions seem plausible, but both fail to do justice to their author's own originality as scientists and to allow for the counterintuitive aspects found especially, but not only, in physics (not their field) since Copernicus. With backward glances to Newton and other great men of the Scientific Revolution, they and their contemporaries prescribed Baconian open-mindedness, painstaking observation, experiment and cautious generalisation as the proper scientific method, though it was by no means always what they practised. Later, the physicist J.J. Thomson (1856–1940) in his memoirs perceptively compared his work on cathode rays in the 1890s with that of his older contemporary the chemist William Crookes (1832–1919): 'In his investigations, he was like an explorer in an unknown country, examining everything that seemed of interest, rather than a traveller wishing to reach some particular place, and regarding the intervening country as something to be rushed through as quickly as possible.'[2] Chemists were voyagers in strange seas, exploring and wondering, like the exuberant Davy, who danced about the laboratory in ecstatic delight on discovering potassium. They then observed and generalised; but Thomson, travelling through the realm of sub-atomic particles in a well-equipped laboratory, had a theory to put to the test of experiment. His vision of real science as the forming and testing of hypotheses that might conflict with common sense became the received view by the later twentieth century, and still is. Nevertheless, there certainly are still people doing science like Crookes', and it does not seem right to view this merely as a natural-history or pre-paradigmatic stage of an adolescent discipline, awkward teenage years to be hurried through. Indeed, Einstein saw scientists as opportunists, using whatever method came to mind or devices to hand; and urged us to focus upon their actions rather than their words. If this seems anarchic as a prescription, these uncertainties and different visions of how to proceed, broadly inductive and deductive, go back to the Scientific Revolution and the great voyages of discovery. In Thomson's day the traditional explorer's task was almost done, but in the 1490s astonished voyagers found countless things of interest to examine. Wide-open eyes, curiosity and the urge to map, collect and classify marked this unsophisticated science.

The great idea lying behind the Scientific Revolution was that insatiable curiosity is a virtue. Our ancestors came to believe that there are more things in heaven and earth than anyone had previously imagined. Like Adam and Eve wiping away their tears after expulsion from Paradise, they saw the world was all before them; and it was wonderful. Human history need not be degeneration from past ages of gold and silver into a grubby and violent age of iron, but could also be progress as wonder and curiosity led into new knowledge nobody had dreamed of. For us, it is a cliché that 'knowledge is power': for Europeans encountering for the first time the works of Greek, Roman and Islamic thinkers, devices coming from as far away as China and the new worlds disclosed by navigators, it was a revelation. They knew that to inherit the Kingdom of Heaven we should be like little children; but here was a promise that wealth, power and longer life here and now would follow when childlike curiosity was made systematic, in science. The biblical curses of hard labour, disease and pain would be mitigated, the spectre of famine exorcised, and military power enhanced. Wonder and delight, standing and staring, were not profitless: science was worth pursuing, for its own sake and for long-term prosperity.

Science has been caricatured as indulging private curiosity at public expense, but curiosity is not enough. The world might be chancy, random, unpredictable, with the past no guide to the future. Curiosity might be, or might have to be, satisfied with the trivial pursuit of dozens of brute facts, unconnected to each other. Another act of faith was required to transform curiosity and wonder into science: faith that we live in a cosmos, an orderly world. Because they believed in God the Creator in whose image humans were made, the Scientific Revolutionaries were confident that they could understand His works, albeit imperfectly. The 'New Philosophers' Bacon and Descartes (both trained lawyers) made laws of nature a focus for scientific endeavour, and nature did indeed seem to obey them. The different versions of scientific method they proposed, and their recipes for progress, resulted from their different understandings of what had gone before, and their own particular interests.

Italy was where it had begun. In the sixteenth century the University of Padua, on the territory of anticlerical, commercial Venice, lacked a powerful theology faculty: the higher degrees available were in law and medicine. Its medical school became famous with the annual public dissections of a felon's corpse carried out by Andreas Vesalius (1514–64), later court physician to Philip II of Spain. Setting his dignity aside, Vesalius got out of his professorial chair from which he was expected to read Galen's greatly respected text, took up the scalpel and displaced the demonstrator at the table. In his book *De Humani Corporis Fabrica* (1543), sumptuously illustrated with engravings from Titian's workshop, he revealed deficiencies in Galen's human anatomy, which had been based largely on apes because human dissection was prohibited.

Aristotle had dissected fish, and took a huge interest in reproduction; for him, explanation involved four causes, which we might prefer to call conditions. There must be appropriate material; there must be some shape, form or blueprint to the process; there must be some reason, purpose or end in view; and there must be some trigger. These were the material, formal, final and efficient causes: when some creature was born, the mother had provided the matter, the father the form, the object was procreation and the trigger coitus. Causation was not simply a chain extending backwards, but more like a network: there are indeed many situations in which causes and conditions are complex and multiple – notably in medicine. In the century before Vesalius, philosophers in Padua had picked up from Averroës a highly deterministic interpretation of Aristotle that was deeply suspect to churchmen, and begun to develop a theory of science that would include restricting 'cause' to efficient causation, with experience and mathematics leading to causal chains and established knowledge, and thence to usefulness. True science would begin with analysis of facts from which a demonstrative, deductive explanation would be derived.

Among the Paduan Aristotelians, Pietro Pomponazzi (1462–1526) embraced a strong determinism and argued that supposed miracles were concatenations of natural events. He also doubted the immortality of the disembodied soul. The most important of these Paduans was Giacomo Zabarella (1533–89), professor of natural philosophy there, who set out to build a coherent system of logic and science. Controversially, in a university where the leading science was the useful art of medicine, he distinguished science from arts as a contemplative rather than useful activity. Among pure sciences he distinguished natural philosophy from metaphysics and mathematics; and he allowed that mixed sciences like astronomy and optics (with empirical, contingent aspects) might be made demonstrative, with fully causal explanations. In the Aristotelian tradition, he downplayed as subservient to physics the mathematics which Galileo, a later Padua professor, would emphasise as the language of nature. Aquinas in that perspective had been taken to separate the natural realm from the supernatural, physics from theology: he had supposed that ultimately theology, queen of the sciences, was the arbiter of truth, but for the Paduans that role went to logic and natural philosophy.

Galileo came from Florence, and saw himself as an exile in Padua from 1592 to 1610, teaching mathematics in that great university. Florence had been the centre of the revival of Platonic philosophy, under the patronage of the Medici family, bankers who became grand dukes and who were also keen promoters of literature, painting and sculpture, and architecture. A different approach to science, and a sympathy not only with mathematics but also with the occult, made this a very different milieu from Padua. For Plato, the real world of ideal forms that we see dimly reflected on the walls of our cave should be the subject of science, and mathematics, too, was a key to them. Marsilio Ficino (1433–99) was the son of the court physician and grew up in the Medici household. In 1439, in a last-ditch attempt to reunite the Western and Eastern Churches in the face of the Turkish threat to Constantinople, Greek delegates came to a church council in Florence; Ficino learned their language, and delighted in the writings of Plato and Neoplatonists including Plotinus (c. 205–70 BC).

Plato had taught in the Academy in Athens, the original groves of academe; and in 1462, under Medici patronage, Ficino was made director of the Florentine Academy. An ordained priest, a great lover of music and a vegetarian, he saw Plato as a gentile prophet to be honoured along with the Hebrew prophets, and as the philosopher par excellence who had demonstrated the immortality of the soul and the dignity of mankind. Ficino became a most important figure, corresponding with humanists all over Europe. His life's work became the translation of Plato's dialogues into Latin, and he gave Plato (and indeed the Neoplatonists) a lasting importance in modern philosophy, notably in Newton's Cambridge. But he was also a great student of astrology, a science in bad odour in church circles because it circumscribed free will; and he put Plato aside for a time while he translated the alchemical and magical texts attributed to Hermes. Science eludes definition, like other complex human activities. Astrology, alchemy and natural magic came to play a powerful role in the early years of the Scientific Revolution, contrasting with and complementing the common sense and logic of Bacon and Descartes. Aristotle and Plato, Pomponazzi and Ficino, bequeathed very different metaphysical frameworks, which their heirs sought to synthesise into a single big picture of a world to be understood and transformed scientifically.

The Magi, guided to Jesus' cradle by a star, might have been kings but were certainly astrologers; and it was generally accepted that phenomena in the heavens illuminated events upon Earth. Comets and new stars were clearly momentous; the idea that the positions of planets might be auspicious goes back to Babylonian times. Early astrology was concerned with states and kings; but by the early years of our era it had been extended more widely, and rules for horoscopes were formalised. If the course of our lives was determined by the stars, then we were not responsible for our actions; but if one believed about the stars, as most do today about genes, that they constrain or limit but do not fix what we do or become, then Christian astrology (as in the tome by William Lilly, 1602–81) becomes a sensible activity.[3] Sceptical Shakespearean characters, baddies like Edmund in *King Lear* and Cassius in *Julius Caesar*, cast doubt on astrology, but most people thought it important: Kepler, as imperial mathematician in Prague, cast horoscopes for Rudolph II (1552–1612), and almanacs were sold in huge numbers, functioning as diaries but also brimming over with predictions. We might suppose that astrology could be tested empirically and refuted, but life is more complicated than that. If an astrologer's prediction turned out wrong, you might change your astrologer; if you reproached him, he might tell you that if you had paid for a more detailed investigation, then the minute influence from Saturn that turned the scale would have been revealed, or that the event showed the need for more research, at some patron's expense. We are familiar, mutatis mutandis, with such arguments in current science, technology and medicine.

Astrology was certainly a reason for promoting astronomy, and in the same way the Hermetic corpus of writings stimulated interest in chemistry. Rudolph and his court were important patrons, and Germany with its ores and tradition of mining and metallurgy was a great centre. The alchemist was an optimist, seeing potential gold where others saw base metal, and self-improvement went closely with the perfecting of material substances. In ancient Egypt, the science was chiefly about making gold, preparing the elixir, alkahest, powder of projection or philosopher's stone that would work the transformation; in China, the prime objective was the elixir of life, which would make our bodies imperishable as gold. From the beginning, alchemy was not only chemistry, but also occult, arcane, a 'mystery' (like crafts, learned by doing them with a master), not to be disclosed to the uninitiated. This made it and its methods very different from the logic and openness that characterise, in principle at least, modern science. We expect scientific illustrations also to be clear, but in alchemical texts sumptuous visual language, enigmatic and symbolic, might accompany texts that were deliberately obscure, sometimes poetic:

> The *Eagle* which aloft doth fly
> See that thou bring to *ground*;
> And give unto the *Snake* some *wings*,
> Which in the *Earth* is found.

> Then in *one Roome* sure *binde* them both,
> To *fight* till they be *dead*;
> And that a *Prince* of *Kingdomes three*
> Of *both them* shall be *bred*.[4]

Nevertheless, this is part of the ancestry of our chemistry, and most if not all of it can be decoded into descriptions of reactions. The spelling 'chymistry' was current for this alchemical science in the sixteenth and seventeenth centuries.

Experimental instructions for the chymistry that grew from the fusion of theory and practice would in due course become prosaic, as in Robert Boyle's (1627–91) writings, and clear. Before him there were – for example, in the *Book of Distillation* (*Buch zu Distillieren*, 1500) of the surgeon Hieronymus Brunschwig (c. 1440–c. 1512), published in English translation by Laurence Andrew about 1530 – woodcuts of apparatus as well as of plants and chymists, so that the reader could get down to work. Although the theme in pictures of alchemists is usually futility and melancholy, chymistry became the first laboratory science, allied to industrial processes in metallurgy, pharmacy and dyeing, where experimental investigation and thinking with fingers, nose and tongue as well as eyes was learned. Chemistry is still a science that begins with and uses all the senses.

Alchemy was perceived both as a practical activity, a promising science, and as a kind of spiritual discipline concerned with perfecting not only the world but also the soul of the practitioner. It was easy to conceive, given Aristotle's notion of elements, Paracelsus' 'tria prima', or classical atomism, that metals might be transformed into one another as their components were transformed, rebalanced or reshuffled. Miners believed that base metals within the earth were gradually being perfected, so it was not implausible to suppose that the process might be catalysed and speeded up in the laboratory. Sulphur because of its golden colour, and mercury, the anomalous liquid metal that forms alloys, 'amalgams', with other metals, were particularly important in alchemical recipes (and in refining silver and gold). Alchemical metaphors were taken up in literature, notably in George Herbert's poem 'The Elixir' (1633), and are, like astrology, a part of the intellectual background of the sixteenth and seventeenth centuries. Boyle and Newton were both devotees, decoding ambiguous recipes, doing experiments and seeking to account for chymical processes and affinities. There were claims of success, even in the eighteenth century: again, it was hard to put alchemy out of bounds just because its central project remained so far unfulfilled. But from the early days there were charlatans out to defraud the credulous (Boyle was a victim), and gradually, like astrology, it had ceased to be reputable about a hundred years before Ernest Rutherford (1871–1937) demonstrated radioactive transmutations, the 'new alchemy', in 1902.

Ficino's interest in natural magic was also widely shared. Hermetic texts included recommendations for amulets to attract favourable influences from planets, as well as incantations to call down gods into their statues and images. To toy with this Faustian material was dangerous in an age living in dread of witches and sorcerers. Nevertheless, contemporaries braved suspicion and argued for the beneficence of natural magic, properly understood and applied. In Naples in 1558 Giambattista della Porta (c.1535–1615) published his book on the subject, translated into English as *Natural Magick* a hundred years later. Della Porta had convened a group investigating the secrets of nature calling themselves the Otiosi, and then became vice-president of the Accademia dei Lincei (Academy of the Lynx-eyed) to which Galileo belonged. The book is a curious mixture, with sections on counterfeiting gold, beautifying women, fishing and cookery, as well as on the generation of animals, magnetism, tempering steel, lenses and other topics:

> They that have been most skilful in dark and hidden points of learning, do call this knowledge the very highest point, and the perfection of

natural Sciences. . . . Others have named it the practical part of natural Philosophy, which produceth her effects by the mutual and fit application of one natural thing unto another.[5]

Natural magic meant getting power over nature, and thus although its occult features seem foreign to science, it was an important component of the Scientific Revolution – whenever we blindly use technology that we do not understand, we are after all still doing natural magic.

In addition to inherited knowledge and craft skills, the creation of modern science thus entailed many things: data collection by explorers, real or metaphorical; Zabarella's Aristotelian logic; Ficino's Platonic openness to music and the big picture; the 'applied' focus of astrologers, alchemists and magicians; the laboratory skills of alchemists; and the stimulus of societies like della Porta's. But there was another crucial thing: mathematics. Platonists respected it, but an appreciation of its power beyond astronomy came chiefly due to Niccolò Fontana Tartaglia (1500–57) of Brescia, who had spent some time in Padua. His name, meaning 'stutterer', was acquired as a boy when his face was slashed by a French soldier in the sack of Brescia and his jaw permanently damaged. Largely self-educated, quarrelsome, finding and losing patrons, in later life he worked in Venice. In 1543 he published the first translation of Euclid into a modern language, and also oversaw the printing of a Latin edition of Archimedes, making his work generally available. Applying mathematics to physics, as in Euclid's analysis of the reflection of light and Archimedes' analysis of levers, balances and densities, brought demonstrative certainty to empirical questions. In his *Nuova Scienza* (1537) Tartaglia applied mathematics to artillery, using a quadrant to measure the elevation of cannons, and finding that 45 degrees gave maximum range. Tartaglia could not prove it because he did not know the projectile's trajectory; but Galileo later demonstrated that it must be so because the shot flies in a parabola. It isn't: for the British Army's 25-pounder it was around 42 degrees, because of air resistance and the gun's irreducible 'jump'. Similarly, Euclid's and Archimedes' proofs apply to perfectly flat surfaces, point masses and weightless struts; but Tartaglia's work marked the beginning of a tradition of geometry in gunnery, leading to the making of instruments for laying guns, for range-finding and for surveying. Artillery became scientific, making castles obsolete and military engineers vital. Physics since Galileo has increasingly been a science of ideal mathematical models adjusted to fit material reality.

We might expect that practitioners of science would be the most respected spokesmen on its methods. But Zabarella and Ficino had no first-hand experience, and neither did the man who came, rather oddly, to be regarded in the English-speaking world as the great authority on how science ought to be done, Francis Bacon (1561–1626). The younger son of an eminent lawyer and a particularly well-educated mother fluent in ancient and modern languages, he went to Cambridge for two years at the age of twelve for a classical humanist education, followed by legal training at Gray's Inn in London. He travelled in France with a diplomatic mission for over two years, but then his father died. Short of money, with support from his uncle Lord Burghley he was in 1581 elected to Parliament. He found a series of distinguished patrons throughout his life, during which he held increasingly important legal positions, one of which involved prose-cuting his former patron, the royal favourite the earl of Essex, in 1601 for rebellion. Knighted in 1603, Bacon became lord chancellor in 1618, and in 1621 a viscount; but in 1622, in the course of a parliamentary controversy about monopolies, he was impeached for taking bribes, pleaded guilty and was forced into retirement. An aristocratic pattern of gift exchange and family honour was starting to be viewed as corruption in a society becoming commercial.

In 1605 Bacon brought out *The Advancement of Learning*, the only philosophical work in English that he published in his lifetime. In it he drew attention to what he called the 'Idols' of the Tribe, the Theatre, the Market Place and Cave, conventions, assumptions, preoccupations and prejudices, which had diverted people into error and he proposed, in their stead, a college of inventors. He wrote that we are the ancients, benefiting from the work of those who have gone before us in the infancy of learning, and called for respectful study of crafts and trades. Science should not mean the contemplation of established truths, but discovery of what was unknown and might become useful. In 1623 he published a Latin version of the book, twice as long; meanwhile, in 1620, he had published *Novum Organum*, the first part of a larger work planned under the title *The Great Instauration* and that would provide a basis for true knowledge leading to power over nature. The book sets out, often in the form of pithy aphorisms, a method of inductive reasoning, beginning with the collection of natural histories. Careful observation and experiment would lead to the exclusion of erroneous explanations, and the formation of true ones.

Bacon has been accused of advocating enumerative induction, the piling-up of cases into generalisations – such as 'all swans are white' – that tell us little and may be falsified by the discovery of a counter-instance. This is not quite fair. In fact, he proposed a form of eliminative induction, and 'crucial experiments' when a crossroads was reached in an investigation and a new direction required. He despised Aristotelian 'final causes' as barren virgins: science was concerned with efficient causation. His legal background shows through, and his idea of subjecting nature to interrogation would mean not only cross-examination but also torture, as was then commonplace in cases involving witchcraft or treason. Science has had a macho image of mastery, even conquest or rape of nature, ever since. Bacon died in 1626, supposedly from catching a cold when stuffing a chicken experimentally with snow to preserve it.

Publishing in Latin shows that Bacon had readers beyond the British Isles in mind. A copy of *Novum Organum* was given to Kepler via Henry Wotton, the British ambassador to the Emperor in Prague. James I, also sent one, reputedly said that, like the peace of God, it passed all understanding, and this sentiment was widely echoed in England where Bacon's contemporaries could not see what he was up to. Bacon's witty *Essays*, published in 1612, have become a classic, and his most influential work was also in English. *New Atlantis*, published posthumously in 1627, was an account of a Utopia, an island in the South Sea peacefully governed through a foundation called Salomon's House, an Academy of Sciences that cultivated useful knowledge and brought a high standard of living to the island's inhabitants.

Baconian science is democratic in that everyone can join in assembling pieces of natural history whatever their status or level of education; but, in accordance with his view of society, it was also hierarchical. In Salomon's House there were degrees: twelve Merchants of Light collected information on foreign countries, three Depredators compiled experiments out of books, three Mystery-men collected them from mechanical arts, three Pioneers or Miners tried new experiments, three Compilers drew up all the results into tables, three Benefactors sought uses from them, three Lamps consulted in devising further and more penetrating experiments for three Inoculators to carry out, from which three Interpreters of Nature then derived further conclusions. There were apprentices; and within the grounds orchards, gardens, parks and lakes, brewhouses and bakehouses, furnaces, dispensaries, textile mills and workshops where the researches were done. The objective was the 'knowledge of causes, and the secret motions of things; and the enlarging of the bounds of human empire, to the effecting of all things possible'.[6]

Bacon's *Sylva Sylvarum*, published with *New Atlantis*, was the work of a Depredator, close to the commonplace books that schoolboys were encouraged to keep where they copied striking passages from what they read. The material is, to the modern eye, very miscellaneous: the author noted that precious stones 'work upon the Spirits of Men, by secret Sympathy and Antipathy', 'in the Art of Memory, that Images Visible work better then other Conceits' and called attention to coffee.[7] By the middle of the century, coffee houses were becoming (as printing houses had long been) an important feature of intellectual life in London and other cities, sites where informal conversation could take place between men of very different social status, business was done, and scientific matters were discussed alongside the exchange of news and gossip.

Bacon could hardly have foreseen that particular development; but he came to be seen as the apostle of applied science and of scientific societies. Curiously, the work of this disgraced courtier and elegant essayist was taken up during the Civil War and Interregnum in Britain (1642–60), when groups of men interested in science and its application met in London and in Oxford. Three men from central Europe, which in contrast to poorly educated England with only two universities was an intellectual powerhouse, played a crucial role in this. They were associated with Paracelsian alchemy, natural magic, projects for the reform of Church, education and society, and apocalyptic speculation, and came to England as exiles fleeing religious wars elsewhere. John Durie (or Dury, 1596–1680) sought to unite the Protestant Churches whose disunion was favouring Catholic

(Habsburg) reconquest of their territories. He emphasised 'practical divinity' as an antidote to what he saw as doctrinal squabbles about words. Comenius (Jan Amos Kamensky, 1592–1670) was a bishop in the Moravian Church. Speaking several languages, he was a great educationalist, pioneering the use of pictures in his textbook *Orbis Pictus* (1659) as visual aids for learning Latin. The 'intelligencer' Samuel Hartlib (1600–70) formed a circle in London to promote education and practical improvement, pooling knowledge and corresponding with the like-minded overseas. All three found in Bacon's writings familiar concerns and became enthusiasts, seeing science as a social enterprise, to be done collectively. This trio, alongside puritans with Geneva connections, now brought England, previously marginal, into the mainstream of European thought.

Overlapping circles were associated with Oxford and Gresham College, an educational institution in the City of London set up by the Elizabethan merchant Sir Thomas Gresham (1519–79). Oxford, formerly the Royalist and high-church capital, was purged of its 'malignants' after Parliament's victory in the Civil War and Oliver Cromwell's future brother-in-law John Wilkins (1614–72) was installed in 1648 as warden of Wadham College. Wilkins was an energetic and powerful figure in reforming the university and making experimental science important there, and in bringing Boyle to Oxford. The phrase 'invisible college' was coined for a group that rarely if ever all met together, but kept in touch through Wilkins and by correspondence; such entities still play a part in science. Wilkins was no narrow puritan, but a liberal, or latitudinarian, in theology, humane, and anxious to emphasise what Christians had in common and to bring them together in one tolerant national Church. Such a broad church, where laymen (and, within limits, clergy) were not committed to particular doctrines, was (and is) a valuable ideal, allowing original thinkers like Newton to remain within it without hypocrisy – in his own eyes at least.

When Charles II was restored to the throne in 1660, Wilkins was one of a group that, after a lecture at Gresham College, resolved to seek a charter for a Royal Society to promote natural knowledge. Later he was one of its secretaries, and in effect its vice-president; and when, in 1665, the Society commissioned Thomas Sprat (1635–1713) of Wadham College and later bishop of Rochester, to write its history, Wilkins was at his elbow.

Sprat asked Abraham Cowley (1618–67), one of the Fellows (who had in 1661 published a proposal for a college for natural philosophy) to write a poem by way of preface. Using the story of the Greek painter Zeuxis (fifth century BC) whose trompe-l'oeil painting had deceived wildlife, Cowley picked up on Bacon's commitment to the novel idea of laws of nature, his empiricism and his prophetic role:

> Bacon at last a mighty Man arose
> Whom a wise King and Nature chose
> Lord Chancellour of both their laws . . .
> Like foolish Birds to painted Grapes we flew;
> He sought and gathered for our use the Tru;
> And when on heaps the chosen Bunches lay,
> He pressed them wisely the Mechanic way . . .
> Bacon, like Moses, led us forth at last
> The barren Wilderness he past,
> Did on the very border stand
> Of the blest promis'd Land,
> And from the Mountains Top of his Exalted Wit,
> Saw it himself, and shew'd us it.[8]

Bacon was also depicted on the frontispiece, gesturing to astronomical instruments and holding a purse. He was the prophet of the new world of practical, organised science which Wilkins and his friends were inaugurating. At a time when prophecy was a matter of deep significance and immense interest, a vision of Bacon as a Moses glimpsing the territory that these Joshuas would enter was resonant indeed, and chimed also with the hopes of those central European exiles. When it became clear that the Royal Society could not expect the kind of lavish royal funding that Bacon had called for, the German Henry Oldenburg (c. 1617/20–77), the other secretary (who had married Durie's daughter) became the most important man in the Society, maintaining foreign correspondence that made it a beacon throughout Europe, and then beginning to publish *Philosophical Transactions*, its journal. This made the Society much more open than Bacon's model, and science a matter of public knowledge.

Bacon's emphasis upon collective work, public science, was thus as important as his inductive procedure, which was not especially novel. He had referred patronisingly and in passing to William Gilbert (1540–1603), the royal physician who had in fact exemplified his inductive methods in a book on magnetism published in 1600. The handsome frontispiece to the second edition of his *de Magnete* (1628) includes a ship, a sailor, magnets and a compass, and also the 'terella' or little Earth, a sphere of lodestone (magnetic iron ore) which Gilbert made for his investigations. Gilbert showed that magnets have poles, and found that the strength of his cigar-shaped lodestones could be much increased by 'arming' them with iron helmets at their ends: from him we have the 'armature' in our dynamos. He put stars of different sizes in the margins of his book when he reported things of especial interest. The attraction of North and South Poles he called 'coition', and believed that this force kept planets in their orbits. Navigators had found compass needles dipping downwards as they went into the far north, and Gilbert with his terella model showed how this would follow from the Earth being a great magnet with its poles below the surface. Navigators had also found that the direction of the needle varied: this Gilbert attributed to the attraction of continents. For Gilbert, magnetism made the world go round, accounting for the Earth's motion around the Sun and revealing life everywhere:

> Miserable were the condition of the stars, abject the lot of the earth, if
> that wonderful dignity of life be denied to them . . . therefore the bodies
> of the globes, as important parts of the universe . . . had a need for souls
> to be united with them, without which there can be neither life, nor
> primary activity, nor motion, nor coalition, nor controlling power, nor
> harmony, nor endeavour, nor sympathy . . . and the whole universe
> would fall into the wretchedest Chaos, the earth in short would be
> vacant, dead, and useless.[9]

Careful experiment and judicious induction had led Gilbert inexorably to these momentous conclusions. Science is diverse.

It also depends upon contingencies. Medicine could lead into science, but for mathematicians in the north, as in Italy, a common path lay through

military engineering. Simon Stevin (1548–1620) from Bruges went north
to join the Dutch in their struggle against Habsburg domination, and
became quartermaster to Maurice of Nassau (1567–1625), Prince of
Orange, who was gradually winning the war. As well as planning fortifications, Stevin worked on the balance and resolution of forces, promoted
decimal fractions (although still with a clumsy notation) and double-entry
bookkeeping, and in lighter vein invented a land yacht, propelled along the
beach by the wind. In 1600 he set up an engineering course at Leiden.
Among his pupils was Isaac Beeckman (1588–1637), trained in medicine
but a keen mathematician, interested in atomism, from a Calvinist family
who, like Stevin, were refugees from the Spanish-controlled southern
Netherlands. Beeckman later set up a technical college in Rotterdam. In
1618 at Breda he met Descartes who, after a Jesuit education at La Flèche,
a law degree at Poitiers and a spell in Paris, was unsure what to do next. He
had joined Maurice's army, as gentlemen did in order to acquire polish and
skills, to travel and, during the longueurs of army life, to think. Beeckman
persuaded Descartes to turn his mind to the mathematical study of nature,
and later introduced him to the work of Galileo.

   In 1621 the truce in the Netherlands was due to expire and more warfare
was expected; but in 1618 what was to be the Thirty Years' War broke out
in a struggle between Protestants and Catholics in Bohemia. Looking for
action, Descartes changed sides and transferred to the duke of Bavaria's
service in 1619. The horrors of that complex and seemingly endless war
appalled contemporaries; it was fought mostly in Germany, for causes as
deeply political as religious. Civilians suffered terribly, and army life became
serious and dangerous; but soldiers went into winter quarters until spring,
and famously, in November 1619 in a warm room near Ulm, Descartes
conceived his plan of remodelling philosophy. Very much a part-time
soldier, he visited Switzerland and Italy, and lived in Paris, where he got to
know Marin Mersenne (1588–1648), a friar who published a famous work,
La Vérité des sciences (The Truth of the Sciences, 1625), against sceptics, and
who maintained a network of scientific correspondence. Descartes took
part in the siege of the Protestant stronghold of La Rochelle in 1628,
but thereafter moved to the Netherlands in order to devote himself to
philosophy in peace and quiet.

He was about to publish his book, boldly titled *Le Monde* (The World), when he heard of Galileo's condemnation in 1633. Descartes was in no danger of the Inquisition in tolerant Protestant Holland, but he had taken the Earth's motion as certain, remained a Catholic and wanted to be read everywhere. So he reorganised his work in the form of an essay, the *Discours de la Méthode* (Discourse on Method, 1637), an intellectual autobiography that preceded a series of treatises on geometry, optics and meteorology that exemplified its method. Geometry was held to be the epitome of deductive logic: from a few axioms, theorems followed inexorably, and could be applied to actual fields and buildings. Algebra in contrast seemed a series of dodges that gave the right answers. Descartes proved that the two were precisely equivalent: geometry could be expressed algebraically using his system of coordinates – what we call the x, y and z axes. Translation between ordinary languages is different: it is never exact. It seemed plausible that Descartes' work might be extended and the same logical certainty found by applying mathematics in different fields, notably optics.

The method involved using systematic doubt to refute those sceptics like Michel Montaigne (1533–92) who, amid the religious and intellectual flux as long-accepted dogmas came under challenge, thought real knowledge was impossible. Descartes found certainty in his famous 'cogito ergo sum' – because he was thinking, he knew he must exist; he then established to his satisfaction that God must necessarily exist also. That provided a guarantee that really clear and distinct ideas were God-given and would be true, and these, verified from time to time by experiment, could be built up into a structured science. His anonymous English translator and disciple wrote that his

> endeavour cannot but be grateful to all Lovers of Learning; for whose benefit I have Englished, and to whom I addresses this Essay, which contains a Method, by the rules whereof we may Shape our better part, Rectifie our Reason, Form our Manners and square our Actions, Adorn our Mindes, and making a diligent Enquiry into Nature, wee may attain to the Knowledge of the Truth, which is the most desirable union in the World.[10]

That this 'New Model of Philosophy' might be carried forward, Descartes invited 'all lettered men' to practise and communicate experiments for the perfection of arts and sciences. He had no time for 'the promises of an Alchymist . . . the predictions of an Astrologer, or . . . the impostures of a Magician'. Aristotelians were 'like the ivie, which seeks to climb no higher than the trees which support it', and should be happy at progress in solving 'divers difficulties of which [Aristotle] says nothing'.

Descartes demonstrated the method in the optics treatise, *la Dioptrique*, where he begins with the insight that because empty space is logically impossible, the world must be full of matter, and the propagation of light is, like the impulse up a blind man's stick, instantaneous: the velocity of light is infinite. Turning to refraction, which makes sticks in water look bent, he found its simple mathematical law, independently of Willebrord Snell (1591–1626) after whom we name the phenomenon. He explained it by suggesting that light goes faster in water than in air. This contradicts his earlier axiom: both could not be right, and so his physics was not as clear and distinct as he had hoped. Elsewhere, he dealt ingeniously with the motion of the Earth: it was swept around the Sun in a vortex of the æther that filled the universe. Both sides in that debate were therefore right: the Earth, like a stick floating downriver, was at rest in its vortex, but also moving around the Sun, carrying the Moon in a mini-vortex.

Extending this mechanical world-view to the whole of nature, Descartes saw animals as automata like clockwork with nervous fluid carrying impulses up to the brain, and animal spirits coming down to inflate the muscles appropriately. By contrast, we had the pineal gland in the brain, where the soul (the 'ghost in the machine') could exercise free choice about our actions. Descartes accepted William Harvey's (1578–1657) work on the circulation of the blood (see below, Chapter 8), but where Harvey made the heart a pump, Descartes made it a pan. Blood entering the hot heart boiled over and bubbled up through the arteries: the power stroke was therefore the expansion, the diastole, of the heart, rather than its contraction, the systole, as Harvey and we believe. Descartes did dissections that seemed to confirm his idea, and prepared his pop-up illustration of the heart to demonstrate what went on. Unfortunately, he accepted an invitation to the court of Queen Kristina of Sweden (1626–89): she demanded early-morning tutorials, and Descartes (who as a boy had been allowed by the Jesuits to lie late in bed, thinking) became ill and died of pneumonia.

Whatever the niceties of method, analogy was and remains the key to scientific explanation. Science and technology were and are enterprises that build upon success, and what has gone well in one branch is worth trying out in another – and that means modelling. Models can be small-scale facsimiles of something, terellas, little ships to impress investors, or the 'Great Model' which Christopher Wren (1632–1723) made of his projected St Paul's Cathedral to convince the clergy that they wanted something of such a radically new form in London. As in Descartes' science, today conceptual models, ideas picked up from mathematics, from devices new or old, or from another speculative domain, are widely used to explain and predict puzzling phenomena. Not everything in the model will fit reality: billiard-ball atoms account for a number of phenomena, but atoms are not hard, coloured spheres. Distinguishing models and hypotheses from well-established facts and theories was and is a major problem for natural philosophers.

Mathematics was a great resource for conceptual models, and during the Scientific Revolution its status rose enormously: what had been viewed as useful training for more fundamental or practical activities became exciting as a route to discovery and explanation. Greek geometry with its structured proofs from axioms continued to be much admired, but applied mathematics occupied a lowly place within the hierarchy of science compared to physics and astronomy: it was to do with possible worlds, a matter of hypotheses and calculations, not concerned with truth about our real world. True knowledge was the business of philosophy and theology. Although Clavius had promoted mathematics in Jesuit education, when Galileo proclaimed that the Book of Nature was written in the language of mathematics, he found himself in hot water. In particular, until Descartes' work, the algebra that had come from the Arabs was suspect because it lacked the rigorous deductive aspects of geometry. But augmented by the numbers the Hindus developed (and the Arabs passed on) and their place-value notation, which replaced sums like MMDCLXXIV-DCCCXIX (impossible without an abacus) with 2674-819, it proved increasingly useful.

Descartes' proof that algebraic propositions could be expressed geometrically, and vice versa, greatly widened the scope of mathematics. The invention of logarithms doubled the life of astronomers (as Kepler put it) by speeding up their calculations, turning multiplication and division into addition and subtraction. The world of Cartesian coordinates expanded at the end of the seventeenth century to include the differential and integral calculus, and the prestige of mathematics (and of quantitative rather than qualitative science) steadily rose accordingly. Uncertainty even yielded to mathematics as games of chance were analysed by early statisticians, and annuities and life insurance were bought and sold. Newton famously wrote: 'hypotheses non fingo' ('I do not feign hypotheses'), and contemporaries came to see his physics as certain; but the view that mathematics was concerned with hypotheses and calculations rather than with under-standing the real world finds an echo much later in unease about quantum theory. The hypothetico-deductive approach of Descartes, for all its uncer-tainties, emerged as a version of scientific method that was rival or comple-mentary to Bacon's inductive fact-collecting, based upon nature's supposed uniformity; it would become the dominant view in our time.

Science meant the pursuit of real knowledge, of truth; but the success of Descartes' incompatible models in accounting for the behaviour of light, Newton and Huygens' disagreement over wave or particle explanations, and Newton's inability to explain the 'action at a distance' involved in gravity, all indicated how in practice those doing science relied upon models, simplified schemata that could be handled and if possible made quantitative. That has remained an important feature of science ever since, as models have been exploited and then abandoned, each seeming a 'truer' approximation of nature than its predecessors. Simple realism was by 1700 not really an option. Scientists used, and continue to use, intellectual models of more or less plausibility and compatibility, following Bacon's dictum that error is better than confusion: mathematics was of great and mounting importance in scientific thinking, but engineering, surgery and chemistry retained many aspects of a craft or art.

Craftsmen's models were therefore equally important, especially of clocks and pumps. Germany was the great centre of clockmaking, but the technology soon spread. By 1284 Exeter Cathedral had a mechanical clock (it needed repair by a bell-founder). The first clocks, driven by weights, kept poor time and needed to be adjusted from a sundial when the weather was fine. By 1392 Wells Cathedral had the splendid clock that still survives, replicating on its face the Earth-centred cosmos, with an hour hand making its full circle in twenty-four hours, a minute hand, the date, phase of the Moon, jousting knights who emerge on the hours, and a figure that strikes the bell. The twenty-four-equal-hour day, beginning at midnight, was coming in. By the seventeenth century, clocks were superb pieces of machinery: even bigger and better than the one at Wells, Strasbourg cathedral's, with its various dials, chimes and moving puppets, was (and still is) a tourist attraction that impressed those who saw it on their Grand Tour.

To its maker in Wells, the clock seemed a little world; to Robert Boyle and his associates, the world seemed like a great clock, a mechanism that we can and must elucidate. Orreries, clockwork models of the Solar System, were constructed for Boyle's cousin the earl of Orrery (1674–1731) and then made for sale. Clocks changed the way people thought of time: the cycle of the seasons, reflected in the agricultural and church calendars, made it natural (despite our aging) to think of the ever-circling years, and to hope that they might bring back the age of gold. Mechanical clocks, although their hands circle their face, went with linear time, an ever-rolling stream flowing equably onwards. Despite our subjective experience of time's circularity (cyclic processes are important in science too), and its going faster and slower, the truth seemed to be that it was linear.

Clockwork, taken apart and put together again in a process that could be dignified as analysis and synthesis, provided one powerful model for science. Pumps were commonplace and provided another, favouring cyclic thinking and a long-suppressed theory of matter. For Harvey, the heart's valves and those in veins, familiar but not fully appreciated, indicated circulation, thus demonstrating the power of a mechanical model or analogy in understanding organisms. Galileo was puzzled that familiar suction pumps, primed with a little water to seal the valve, will not lift water more than 30 feet (9 m), and his disciple Evangelista Torricelli (1608–47) inferred that we live at the bottom of a great ocean of air pressing down upon us. Boyle got Robert Hooke to make him an air pump, and they experimented publicly with it, showing, for example, that a bell in the receiver became inaudible: there really was a vacuum in there. For them and their fellow empiricists, the pump established that Democritus (c. 460–c. 370 BC) in ancient Greece had been right: the world was composed of atoms and void. This 'corpuscular' view of nature came to prevail by the eighteenth century.

More abstract analogies arose with law, a matter of huge importance in the sixteenth and seventeenth centuries as the Reformation led to wars and civil wars, and to the trial and execution in the Netherlands of a grand pensionary (prime minister) and in Britain of a monarch. Law had been crucial to the Roman Republic and Empire, and their law was the model for that of the Church, and for most European civil law. Hugo Grotius (1583–1645) developed the idea of natural law, binding all humanity; while in Protestant England 'common law' based upon tradition prevailed. Bacon and his contemporaries picked up and developed the idea that things were subject to laws imposed by God: a notion that became central to science. To contemporaries of Galileo, Boyle and Newton, such God-given, universal laws of nature were far superior to human laws, and were sought because they seemed definite and absolute. Once the law had been found, the inevitably hypothetical and provisional scaffolding of explanation could be removed from the edifice of science.

Descartes' power, confidence and range are dazzling, and his ambition to be seen as the founder of modern philosophy has come true. He hoped to be able to do the reconstruction of knowledge on his own, but had to admit that it would be the work of generations yet to come. He outlined a system of knowledge that could replace Aristotle's, and his science entered courses in universities: the Cartesian *Traité de physique* (*Treatise on Physics*) by Jacques Rohault (1618–72) became a standard textbook, notably in Newton's Cambridge. It was translated into English, and in later editions given footnotes by the Newtonian Samuel Clarke (1675–1729) that often contradict the text. In the seventeenth century, scientific 'explorers' would follow the inductive method of Bacon, and 'travellers' the hypothetico-deductive route of Descartes. Both routes promised a certainty that we now know was illusory: science is deeply provisional. Even though there is an expanding core of well-established data, its interpretation and significance will change with time and fresh discoveries. Bacon and Descartes were lawyers and sons of lawyers; and whereas inductive and deductive proofs seem a scientific ideal, legal proof beyond reasonable doubt is in fact the best that can really be hoped for in the empirical world. Although a deep respect for evidence is the first requirement of every scientist, science like all human affairs demands judgement about probabilities and possibilities depending upon world-views, and upon fashion. Four centuries after Bacon and Descartes, concern about method is more acute nowadays in the more self-conscious world of the social sciences than in the long-established natural sciences, where it is taken for granted – though the value and very existence of method have recently been challenged by the maverick philosopher Paul Feyerabend (1924–94).

Checking calculations, repeating experiments and the ethos of public knowledge should make scientific knowledge sure; but fraud is possible, has sometimes happened in science and, while ideally it would be speedily revealed, can distort understanding for years. It is not always straightforward. Galileo claimed that his experiments rolling balls down slopes to test the law of falling bodies were always spot-on: that cannot have been strictly true. Boyle announcing the law that bears his name, published raw data. Galileo, like physicists since, was testing a formula he had deduced; Boyle was working inductively and generalising from his data, and that still happens too. But scientists have not always been as scrupulous as Boyle in handling and publishing their results. As with all crime, there are degrees of fraud in science, and even perhaps white lies. The 'trimmer' tidies the results in accordance with confident conceptions of what the law must be, omitting 'bad' readings; the 'cook' fakes, as all students must have done on occasion, working back from a desired outcome to the readings that would have produced it. Because of instrumental imperfections, and because research is always being done at the limits of knowledge, creative imagination is often required, and can lead astray: some preformationists with microscopes saw spermatozoa as tailed homunculi, and Galileo's depiction of the Moon seen through his telescope was overdetailed. Seventeenth-century men of science were well aware of the importance of caution and of using the best instruments, and might, like Hooke and Antony van Leeuwenhoek (1632–1723), become highly skilled in making their own, keeping one step ahead of the instrument-makers, and even sending a prism or microscopes, along with instructions, to help others understand and replicate their observations. Nevertheless, claiming priority entails speedy publication, which as well as keeping science moving can result in the propagation of serious error.

Gentlemen of science, independently wealthy like Boyle and Henry Cavendish (1731–1810), might be relaxed about priority and dilatory about publishing; but most people had and have to 'work, finish, publish', as Michael Faraday (1791–1867) put it. Their originality and priority are their stock-in-trade, vital to them in making their way; and the introduction of scientific journals, giving regular periodic publication, on top of meetings of academies and societies where discoveries could be announced and demonstrated, were crucial for them. Fame and merit remain very loosely associated, even in the sciences, especially because we like to attach a discovery to a name. We are familiar with 'breakthroughs' that come to nothing, and these are not new: claims for cures, for making gold, for perpetual motion, for finding longitude happened in the Scientific Revolution too.

Science is a collective enterprise, a band of brothers and sisters, in many respects, but it is competitive: for resources between different disciplines and institutions, and for recognition, institutional and individual. Poets, painters and musicians, though subject to rivalries, the carping of critics and the vagaries of taste, can delight in being part of a movement, an avant-garde; scientists can feel that too, but know also that others working in the same field are competitors, that almost exactly simultaneous discovery happens and that there are no second prizes in their objective world. Even the most theoretical and abstract science is closely dependent upon technology in the form of equipment and apparatus; and science is supported at huge expense directly by taxpayers and indirectly by consumers because it brings power, mastery, utility, convenience and wealth. Mandarins try to separate pure and applied science, but this has rarely been easy and seems ultimately futile: our technology is the fruit of science, and by its fruits one judges the tree.

What we have seen here are aspects of science, visions of how to proceed, and from them we can discern factors that were crucial in bringing the Scientific Revolution about and keeping it going: in Vesalius, the critical spirit that questions authority and looks harder; in Pomponazzi, the importance of logic and scepticism; in Ficino's astrology, prediction; in Paracelsus' alchemy, experiment; in Tartaglia, Galileo and Descartes, mathematics; in della Porta, Mersenne, Hartlib, Wilkins and Oldenburg, societies and academies, public knowledge; in Bacon and Descartes, laws of nature, publicity and celebration; in Gilbert and Wren, models; in Hooke and Boyle, machines and then, with them and Harvey, conceptual models. Science works not by following an infallible royal road, but through curiosity and imagination, criticism and rigorous testing, and the building of a community. It is rooted in the societies in which it is a force for change.

The history of science is thus contingent and its development a feature of the culture of times and places; but that has not stopped the philosophical-minded from generalising the polar opposition of Baconian and Cartesian methods. Thus Pierre Duhem (1861–1916) contrasted the broad and shallow English mind with the deep and penetrating French one, though seeing both as necessary for scientific progress. This was not simply a curious example of nationalism, for Duhem placed Napoleon amid the English and Newton among the French. In a different dichotomy, the cultural historian Henry Buckle (1821–62) contrasted the plodding, inductive masculine intellect with the leaping and deductive feminine that he had observed in his own mother. He believed that only when both were involved could science develop properly – but his feminist case was weakened by his choice of Newton's as the example of the female mind. There is no doubt that there are different kinds of mind, different styles of thinking, appropriate to different sciences at different times; another distinction is between those who seize upon the big picture and look for connections between things, and those who specialise. We shall meet all sorts in this history. Seeing how science has transformed itself, and how provisional it is, we may think that there is no foolproof means of discovery, no road straight to certainty. But our ancestors, entering a new world, were more optimistic, and questions of method and satisfactory explanation were as acute in seventeenth-century astronomy as they now are in 'evidence-based' medicine. And it is to astronomy, often portrayed as the crucial element in the Scientific Revolution, that we now turn.

CHAPTER 4

LOOKING UP TO HEAVEN
MATHEMATICS AND TELESCOPES

PERFECTING THE CALENDAR WAS one long-standing incentive to study astronomy in the West, for Bede, Sacrobosco, Regiomontanus, Clavius and young Copernicus; but that was settled by the Gregorian reform. The new spur was navigation, as sailors went way out of sight of land on ocean voyages that became a metaphor for science. The revelation of the globe to and by the navigators brought renewed value to astronomy, which flourished in the wake of practical concerns before becoming an exemplary exact science. Trade, conquest, settlement and exploitation of resources lay behind the investment in observatories that was a feature of the Scientific Revolution, notably in Hveen, Paris and Greenwich. Starting out from the Canary Islands, Madeira and the Azores, Iberian sailors had ventured further and further into the unknown. They had to determine where they were, and plotted their course reckoning how fast they'd gone and in which direction, and checking the result by taking celestial observations that should have given them their latitude but that were tricky to arrive at on a heaving deck or a cloudy night. Longitude was very difficult, and could only be determined by making careful observations ashore. Then, on landing, observation of the heavens could tell them where on Earth they were. Here was work for astronomers.

Copernicus hoped that moving the centre from the Earth to the Sun would simplify calculations, but, using the same data and similar epicycle constructions, he found that to get the same accuracy he needed as much geometrical complexity as Ptolemy. What contemporaries liked about Copernicus' theory was that it was coherent and conceptually simpler. Watching Venus, Mercury and the Sun, Ptolemaic astronomers could not say which was nearest: they all took a year on their orbit round the Earth. Usually the Sun was deemed furthest out, making it halfway between the Earth and the stars: the Moon, Mercury and Venus were within its orbit, and Mars, Jupiter and Saturn beyond it. Whereas for Ptolemy the orbit of each planet was computed as a separate problem, for Copernicus there was a Solar System: Mercury went round the Sun quickest, so must be nearest, then Venus, Earth, Mars and so on. Awkwardly for him, though, the Moon manifestly circled the Earth, so he required two centres instead of one.

Ptolemy adopted the view, or model, that planets move in epicycles: that is, they move in a circle about a centre on the circumference of a big circle, the deferent, itself moving around the Earth. Epicycles are necessary because the planets do not move around us regularly: as they all travel in the same plane (the ecliptic) against the background of the fixed stars, we see them speed up and slow down in their passage, and from time to time stop and go backwards for a bit. These stations and retrogradations, in which they are brightest, happen predictably: and for Ptolemy, the planet, going round the Earth on its main circle (the 'deferent'), revolves also in its epicycle; when the two motions are opposed it comes to a halt and goes backwards. For Copernicus, though, retrogressions were simply the effect of planets' motions round the Sun. Thus Mars' year is twice Earth's, so we regularly overtake it on the inside; as we pass it, it seems to go backwards, and as we are then nearest, it is brightest. Copernicus still needed epicycles and other devices to get exact agreement, but his account was elegant.

Ptolemy was content that his model 'saved appearances'; Copernicus believed that his system was the true one, but he and his disciples faced major problems in proving it. The first was parallax. Out walking, we notice that the angle between two distant trees increases as we get nearer: the copse opens up. This is the effect of parallax. The ancients knew that the angle between any two fixed stars stayed the same wherever they viewed it from on Earth. The absence of parallax meant that the stars were so far away that terrestrial distances were insignificant: the Earth was but a point in comparison to the heavens, as Ptolemy put it. But if the Earth were moving in an orbit millions of miles across, then we should still see some difference in those angles after six months. Copernicus knew that we don't. Instead of taking this as falsifying his promising hypothesis, he suggested that the stars were much, much further away than anybody had thought, and that the available instruments were too crude to measure the minute changes. The Earth's very orbit was insignificant. He was right, and three hundred years later parallax was detected; but in his time his vast, useless voids in space seemed absurd, perhaps terrifying. Although he did not need a sphere of fixed stars (because they don't revolve around the Earth), he hung on to it: but a convert, the practical mathematician Thomas Digges (c.1546–c.1595), editing his father's almanac *Prognosticon* in 1576, filled the page around the Copernican Solar System with stars extending indefinitely, in perhaps an infinite universe.

Copernicus had defied not only scientific method, but also the physics of his day. By making Earth a planet, he abolished the distinction between the terrestrial realm, where four elements go up or down, and the celestial, where quintessence goes round and round. Gravity had been explained in terms of heavy matter falling towards its natural place, the centre of the universe where Earth was. If the Sun were the centre, then everything should fall towards it. A different, ad hoc theory was required, in which perhaps earthy matter went towards the centre of the Earth, solar matter towards the Sun, and so on: the 'Moon dust', eventually brought back to us by NASA, would shoot upwards to get back where it belonged. Moreover, if the Earth were moving in orbit, one would expect that we should see the effects: lighter things would be left behind, and there would be tremendous winds. If it were also spinning on its axis, these would be more pronounced: the circumference of the Earth being 24,000 miles (38,400 km), a point on the equator would be travelling at a thousand miles an hour (1,600 km/h). If a bird let go of a tree it would be whizzed away at impossibly high speed. Keeping the Earth's axis pointing the same way all round its orbit was another problem, requiring another circle.

There were thus good reasons, quite apart from worries that it contradicted the Bible, for disbelieving Copernicus' claim, but the book (*De revolutionibus orbium coelestium*, 1543) was taken seriously, and many copies survive, suggesting that the aesthetic appeal of his system must have been an overriding attraction. It came with a dedication to the pope; but a misleading and unauthorised preface by the publisher, Andreas Osiander (1498–1552), a Lutheran theologian of Nuremberg, indicated that it embodied no more than a mathematical hypothesis to simplify calculations. This unsigned preface was generally supposed to be by Copernicus, who was on his deathbed in far-off Frauenburg when the book came out in 1543, and it temporarily disarmed possible critics in the Churches.

Copernicus, no great observer, relied upon data he inherited, accumulated over centuries of scribal copying and of variable initial accuracy; some were illusory because of poor observation or miscopying of texts. A prominent Danish nobleman, Tycho Brahe was attracted to astronomy because of its potential for predictions but also because of its present inaccuracy. In 1572 he saw the new star in Cassiopeia that also excited Digges: astonishingly, its lack of parallax showed that it was in the supposedly perfect and unchanging heavens. Tycho persuaded King Frederik II of Denmark (1534–88) to grant him the island of Hveen and endow an observatory he called Uraniborg (begun 1576) with magnificent equipment and a printing press: astronomy was the big science of its day. There, for twenty-one years, with various assistants he followed the Moon and planets right round their orbits and not merely at interesting points. A scrupulous observer, he estimated and allowed for the errors on his enormous instruments and for refraction, improved their scales and the design of their sights, and reached the limit of the naked eye: better than half a minute of arc. He proved that the comet of 1577 was beyond the Moon, and therefore a heavenly body rather than an atmospheric phenomenon. Unable to accept Copernicus' theory, he proposed a geometrically identical variant: the Sun circled the stationary Earth, carrying all the planets with it. This Tychonic system seemed a best-of-both-worlds compromise.

In 1597 he fell out with the new king, Kristian IV (1577–1648), and departed in a huff for Prague, where he was welcomed by the Holy Roman emperor, Rudolph II, patron of artists, scribes, astrologers and alchemists, and owner of a wonderful cabinet of curiosities. Rudolph gave him a castle, and there, in 1602, his *Astronomiæ Instauratæ Mechanica*, with magnificent plates and descriptions of Uraniborg and its equipment, was published. By then Tycho himself was dead. He is buried in Prague, and on his effigy can be seen the line showing where his nose had been cut off in a duel: disconcertingly its silver replacement could be taken off and polished up as necessary. The book was edited by Kepler, his successor as imperial mathematician, who though a Protestant had come to Habsburg Prague to work with Tycho and get his hands on Tycho's data.

Kepler was a Copernican, who had already devised an ingenious geometrical construction for the Solar System in which the orbits fitted neatly into regular solids; when this failed to square with Tycho's accurate observations, he resolutely dropped it and tried other ideas. He described his labours in his handsome *Astronomia Nova* (1609), in which Tycho's name appears as prominently as his own on the title-page: Kepler was a modest man. The book is hard to navigate now, because Kepler included what he knew were his false steps on the way to arriving at the laws that bear his name: that a planet moves about the Sun in an ellipse with the Sun at one focus, sweeping out equal areas in equal time and thus speeding up when it's nearer the Sun. Kepler accounted for the laws by adopting William Gilbert's magnetic ideas. He believed that mathematical harmonies must govern the world, but was boldly prepared to abandon Plato's programme of using perfect circles to account for planetary motions. The labour of computing possible orbits from observations was immense; as noted in the previous chapter, he later welcomed the invention of logarithms by John Napier (1550–1617), published in 1614. To help the reader navigate through 350 folio pages of calculations, some of which he knew were fruitless, Kepler included a fold-out 'Synopsis'; but even so, and despite the printer having decorated the diagrams with flowers, the book was hard going, and the laws not easy to spot.

Rudolph's court was a centre of patronage where money, modernity and tolerance went together. But in 1618 the Thirty Years' War began: central Europe was its battleground, the zest for novelty there was quenched, and the Netherlands, France and England took up the torch. Even earlier, Kepler's life became difficult: his wife and child died in 1611, and Rudolph's throne was usurped by his brother Matthias who neglected to pay Kepler's salary. During the war he moved to Linz, to Ulm and then to Silesia, before making a final journey in search of his backpay to Regensburg (Ratisbon), where the Imperial Diet used to meet and where he died. Meanwhile, he had had to get his mother cleared of witchcraft. Nevertheless, he published the *Rudolphine Tables* (1627), systematising the work of Tycho (whose name this time appears larger than his own on the title-page) and featuring a magnificent frontispiece illustrating allegorically the history of astronomy, and in particular the role of Hveen and its printing press. Kepler worked on optics and vision, realising that the image on our retina is upside down, and on ice crystals; but he saw as his greatest triumph *Harmonices Mundi* (1619), in which he printed the music of the spheres, computed from ratios of the planets' speeds and distances. In the course of this, he worked out by crunching numbers what we call his third law: that the squares of the periodic times of the planets are proportional to the cubes of their mean distances from the Sun – thus linking together the Solar System in harmonious unity.

Kepler's works might well have gradually convinced astronomers that Copernicus was right: in Wilkins' Oxford and at Gresham College, Wren pioneeringly lectured on the elliptical orbits until the Fire of London (1666) transformed him from the star of the early Royal Society into a full-time architect. But meanwhile Galileo (who never bothered with Kepler's laws) heard in Venice – a centre of information, printing and glassmaking – about the Dutch invention of a spyglass. In 1609 he made one himself and looked upwards: he was astonished. Science would never be the same again. This new instrument, making visible what had previously never been seen, proved revolutionary: its successors continue to transform science, medicine and technology. An excellent observer, Galileo saw how stars, of which there were thousands, looked different from planets that showed as discs. He spotted how four little stars near Jupiter were actually moving around it, like a miniature solar system. He saw craters and mountains on the Moon, and from their shadows estimated their heights: it was manifest to him that here was an earthy body, rather than a perfect globe of quintessence. In 1610 he published *Sidereus Nuncius* (The Starry Messenger), with woodcuts of the Moon, and became instantly famous. The Venetians, convinced of the usefulness of what was soon called the telescope, offered to double his salary: but Galileo had an eye on different patronage. He named the four moons of Jupiter the Medicean Stars; and Cosimo de'

Medici (1590–1621), grand duke of Florence, duly offered him the position of court philosopher. Galileo had been a closet Copernican, but now emerged from his place of intellectual hiding, knowing that our Moon was not unique, because Jupiter had moons too (and was the centre of a miniature solar system), and that Venus had phases like the Moon, which was only possible if it circled the Sun. Believing firmly that the heavens and the Earth must follow the same laws, he sought a new physics that would work on a moving planet.

Kepler, who rejected Tycho's system because it seemed dynamically absurd that everything should circle the tiny Earth, was sent a copy of the book and confirmed its findings. So did Clavius, the veteran Jesuit astronomer based in Rome who had supervised the Gregorian calendar reform. Galileo was elected to the Accademia dei Lincei. A court philosopher was expected to debate, and Galileo developed a style of witty invective in his native Tuscan that might carry the day but did not always win friends. Despite Galileo's lobbying, in 1616 the Roman Inquisition condemned the Copernican system as probably heretical because it contradicted Scripture, and Cardinal Robert Bellarmine (1542–1621) admonished Galileo not to defend it in public. A learned Jesuit, he admired Galileo, and remarked that if Copernicus should be proved right, the Church would duly reinterpret those passages of the Bible that seemed to contradict him. These were not, after all, crucial to central tenets of Christianity, but Catholics had become sensitive to Protestant charges that they played fast and loose with Scripture. Galileo hoped to prove the Earth's motion by an argument from the tides, but was faced with the problem that probability beyond reasonable doubt is the best we ever find in science, and that good theories must, to be useful, anticipate much of the evidence that later confirms them.

In 1623 Galileo published *The Assayer* (Il Saggiatore), an entertaining polemical essay about scientific method much enjoyed by his friend Maffeo Barberini (1568–1644), who had just been elected Pope Urban VIII. But it was deeply offensive to its target, Horatio Grassi (1583–1654), a Jesuit astronomer who had (using the pen name Sarsi) disagreed with Galileo about the comets of 1618. Galileo's ridicule of Sarsi made him enemies among the Jesuits. The Inquisition's investigation had originally been set going by smarting Dominican adversaries – to provoke both Jesuits and

Dominicans, powerful teaching orders often at loggerheads within the Church, was rash. Galileo was not, like Kepler, humble. He forthrightly declared that the Book of Nature was written in the language of mathematics, praised Copernicus for defying common sense and boldly revived from the ancient atomists (notorious atheists) the idea that secondary qualities (colours, tastes and smells) were the result of arrangements of particles that had only the primary qualities of weight and shape. The task of science would be to demonstrate that just as tickling 'belongs entirely to us and not to the feather ... I believe that many qualities which we come to attribute to natural bodies, such as tastes, odors, colors, and other things may be of similar and no more solid existence'.[1] As well as mathematics, experiment or experience (for the two terms were not yet clearly distinguished) to test hypotheses would be essential, though, like a philosopher arguing a case through examples, Galileo sometimes employed armchair thought-experiments rather than the hands-on kind. Galileo went to Rome in 1624, when the pope sanctioned him to compare the Ptolemaic and Copernican systems in a book, and in 1632 he published his *Dialogue on the Two Great Systems of the World* (*Dialogo sopra i due Massimi Sistemi del Mondo*) in Florence, where it had been approved by the local office of the Inquisition.

The three speakers in the book meet over four days in Venice: Salviati, representing Galileo, Sagredo, his pupil, and Simplicio, a rather dim conservative. Salviati explains how, as mentioned above, Galileo had devised his new physics for a moving Earth. On a smoothly moving ship, a stone dropped from the top of the mast falls vertically: unless the ship is speeding up or slowing down, everything on it shares its motion and behaves just as it would were the ship at rest – and thus we and everything else partake of the Earth's motion. Falling bodies accelerate following the rule devised by the Merton mathematicians (though Galileo does not mention them). And were a ball to be rolled along a frictionless surface it would go on for ever: it is friction and air resistance that bring things to a halt. Given all this, is it possible to demonstrate the Earth's motion? One day Simplicio is late because his gondola is grounded by the ebbing tide; he wonders why this happens. Luckily for the Venetians, tides there are small; Galileo believed they happened once a day, there and everywhere. Bede, living by the ocean, had known better. Galileo inferred that the tides were the result of the Earth's motion: he rejected the connection with the Moon, because it seemed occult. For him, balls rolled along smooth surfaces, like Francis Drake's (c.1540–96) in his game of bowls awaiting the Spanish Armada, would roll right round the Earth and come back: the Moon is just such a ball, rolling round us thousands of miles up where there is no friction or air resistance to stop it. Planets similarly moved round the Sun in circles, mildly distorted by minor causes. Here Galileo's physics differs from ours: for Descartes, Newton and us, bodies continue in a straight line unless a force acts upon them, and there must therefore be something that keeps the Moon in orbit.

Salviati compared Copernicus' system to Ptolemy's and ignored Tycho's, which was favoured among conservatives and which he could not refute. He had replaced Aristotle's with a new physics, in which the Earth's motion makes no difference, and had vastly increased Copernicus' plausibility through observations and reasoning. The amiable and almost convinced Simplicio was left to make the point that we could never be sure that our science disclosed just how God had actually made the world. This was something that the pope had said to Galileo, when authorising him informally to publish what he expected to be a comparison of the two systems for a learned audience; it was unwise to put his words into Simplicio's mouth. Galileo must have known he was on thin ice, but did not expect the fury he provoked. As happens with complex events, there have been many possible explanations for his condemnation, which in the long run did such damage to the Roman Catholic Church. It might have been catastrophic for men of science, bringing another promising beginning to a premature close as had happened in Averroës' time; but in the event it was not. There was too much momentum already, too many factors promoting science in the broadest sense, and the complex warring world of the seventeenth century proved a hotbed in which it flourished; but that did not save Galileo from the pope's wrath.

Patronage is an important key in Galileo's case. Popes tended to be elderly and their tenure short: but Urban reigned for over twenty years, 1623–44. He made two nephews cardinals, which was going rather far: grievances among those excluded from his patronage grew over time.

Moreover, in the Thirty Years' War, which was supposedly an affair of Protestants against Catholics, he supported the French and their Protestant allies against the overmighty Habsburgs, who were outraged. Urban was therefore under pressure from Spain or Austria, and the classic response of princes is to sacrifice a favourite or client, as Charles I (1600–49) did with Archbishop Laud (1573–1645) a little later. The Venetians in such a case would have been capable of standing up against the offended and belea-guered pope if he had remained there; Galileo had prominent friends and allies in Rome as well as the grand duke in Florence, but they could only mitigate his punishment. Though an old man in poor health, he was summoned to Rome. The book had not been the anticipated dry Latin compilation, but persuasive and accessible, in Tuscan Italian: it conformed to the letter but not the spirit of what had been authorised. The Inquisition did not make formal charges or presume innocence as a modern court might do. It was not clear whether Galileo had been cautioned by Bellarmine, or merely informed; but most expediently a document was found (or prepared) which established that he had received an order and disobeyed it. Bowing to the inevitable, he made a public recantation in a show trial, and was sentenced to house arrest for life in his villa outside Florence. Jesuits at the Chinese court, who had been teaching Galileo's discoveries as an example of Christian advances in knowledge, had embar-rassingly to change course.

The restrictions on Galileo were stringent and never relaxed, even though, like a Soviet dissident in the twentieth century, he became a kind of tourist attraction; the Florentines were later prohibited from giving him a civic funeral. His distinguished visitors included the crown prince of Poland and John Milton, who later wrote of 'strange Parallax' and 'Optic skill of vision, multiplied ... by glass of Telescope'.[2] Galileo remarked bitterly that the innocent can never be pardoned. Nevertheless, despite encroaching blindness, he wrote a *Discourse concerning Two New Sciences* (*Discorsi e dimostrazioni matematiche intorno a due nuove scienze*), which was published in Protestant Leiden in 1638. This has the same three characters as the *Dialogue*, but it is a much more formal work, a tutorial rather than a conversation. It was dedicated to his friend the comte de Noailles (c.1613–78), French ambassador in Rome, and Galileo, more prudent in his old age, pretended that he had lent him the manuscript and then found to his surprise that it was being published. It is in this book that Galileo set out his dynamics, an inertial physics in which the parabolic path of projectiles was demonstrated, with illustrations and geometrical diagrams. Although the Inquisition had forbidden the publication of his writings, this book and Galileo's earlier ones were never actually placed on the Index: presumably no prominent complaints were made about it, and the Inquisition was reactive rather than proactive.

Though they did not stop him doing physics, Galileo's trial and conviction were a hugely important event, the ruthless suppression of one of the great men of the age, and have justly entered the realm of myth, significant stories that point beyond themselves. While often portrayed as a battle in a long-running war between science and religion, it is better seen as an episode in the gradual seizure by laypeople of roles, authority and power previously held by the clergy. That is a somewhat different war, and by no means over everywhere even today: but right through the Scientific Revolution what Samuel Taylor Coleridge would call a 'clerisy' of learned professionals was claiming (in the wake of Galileo) philosophical, moral and scientific authority alongside, and independently of, the clergy. Separate spheres, intellectually speaking, were to become the order of the day; and Galileo's once-shocking assertion that astronomers were the proper judges of astronomical claims has come to seem incontestable. The irrepressible Galileo had included references to St Augustine (for whom Genesis 1 could not literally be true) on scriptural interpretation and the need for flexibility in the face of evidence. He had also permitted himself a certain flippancy about the story that God stopped the Sun in its tracks so that Joshua could finish off the destruction of Israel's enemies: he noted that if the Sun were travelling in an epicyclic path, it would actually have to speed up in order to seem to stand still. It was outrageous to Inquisitors that witty laymen should claim to teach them how to read the Bible; but the new breed of scientific sages would claim that right, in addition to pontificating upon scientific matters. Luther had laughed on hearing about Copernicus' ideas; but papal condemnation meant that learned Protestants ceased to doubt that the Earth moved – as did Catholics in France, where papal bulls and Inquisition rulings only applied if the king allowed them.

Galileo was not the only one to point a telescope to the skies. Kepler told him that Rudolph II thought he saw through his spyglass an image of Italy projected upon the Moon's surface. And Thomas Harriot (c.1560–1621) had in London in July 1609 (the very same time as Galileo) made or acquired a similar instrument and espied the moons of Jupiter and the craters on the Moon; but he was unfortunate in his patrons. His first, Walter Raleigh (1552–1618), under whose auspices he had trained sea captains in navigation and whom he had accompanied on the voyage to set up the Virginia colony at Roanoke in 1585, was charged with treason and confined in the Tower of London: Harriot was suspected (wrongly) of infecting him with atheism. Raleigh's friend Henry Percy (1564–1632), the 'Wizard Earl' of Northumberland, took Harriot on, but then he was accused of complicity in the Gunpowder Plot of 1605 to blow up King James and the Houses of Parliament, fined and sent to the Tower also. As the associate and (possibly) evil genius of both, Harriot was under suspicion: it was rumoured that he had cast the king's horoscope to see if a sticky end was imminent, and he was well advised to keep his head down. To a reserved person, as Harriot seems to have been, this was less of a blow than it would have been to the exuberant Galileo; but be that as it may, he published very little apart from his *Briefe and True Report of the New Found Land of Virginia* (1588), with its fascinating and sympathetic account of the Native Americans' culture. Outside a small circle, his highly original work in optics and in algebra was little known to contemporaries, and was only revealed by twentieth-century research. Patronage had brought him a life with leisure to pursue the science he loved – but with drawbacks that to many would have been irksome. When discoveries are made simultaneously, as they often are, honour and reputation (and criticism) go to whoever manages to blow the trumpet louder, as Galileo did in the case of his spyglass work.

Problems with patronage, suspicions of casting royal horoscopes and accusations of necromancy had earlier attended John Dee, sponsor of the young Digges and philosopher to Queen Elizabeth I (1533–1603) from 1570, some of whose instruments may have passed to Harriot. Dee's career began glitteringly: after studying at St John's College, Cambridge, he became a Fellow of the newly founded neighbouring Trinity College, and visited Louvain and Antwerp. But on his return to England, in 1555 under the Catholic Queen Mary (1516–58), he was arrested as a magician or 'conjuror'. Released, he lavished praise on Copernicus in 1556, and in 1570 wrote a 'mathematical preface' to an English translation of Euclid that became a classic of scientific method. He promoted navigation and discovery, and devised stage machinery; in addition, along with a library and instruments for astronomy and surveying, he possessed an obsidian mirror and various crystal balls which he used for divination, the boundaries of his science being fluid. In sessions with his disciple Edward Kelley (1555–97) from 1583, he held conversations with a number of angels. Invited to Poland, he went on to Bohemia, where in Prague in 1584 he picked up a medical doctorate and had an audience with Emperor Rudolph. Back in England from 1589, with a large family to support, he found his kind of natural magic and forecasting losing favour, especially after the accession in 1603 of James I. James' mother's voyage to Scotland had supposedly been endangered by witch-induced tempests (picked up by Shakespeare in *Macbeth*), and he was much agitated about what he saw as the proliferation of witchcraft. Dee died in poverty, his reputation as a natural philosopher and mathematician much damaged by those angelic conversations and the suspicion that he was a wizard.

In the Netherlands, the very different kind of thinker Descartes reorganised his material after hearing about Galileo's condemnation in 1633, and his firmly mechanical physics and physiology proved highly attractive to students. His world had its x, y and z axes stretching to infinity, and for him a body's inertia would carry it in a straight line. So Galileo's belief that the Moon just rolled effortlessly round the Earth, and the Earth round the Sun, would not do: Descartes believed that space was full of subtle æthereal matter, and invoked whirlpools or vortices of it to carry planets round. The Earth, swirled round in the Sun's vortex, was at the centre of a smaller vortex that carried the Moon. Bodies that had got loose from a vortex were whirled hither and thither as comets until recaptured by another vortex. This was a satisfyingly comprehensible replacement for the nests of spheres and epicycles of previous generations, and duly prevailed widely; but how it worked in detail Descartes could not demonstrate.

To get the new astronomy across, authors resorted to science fiction. Picking up on Utopian literature, they imagined space travel and peopled heavenly bodies with intelligent beings. Kepler's *Dream* (*Somnus*) was published posthumously in Latin in 1634, having circulated in manuscript in his lifetime since 1611. During an eclipse and thus shielded from the power of the Sun, his traveller was whizzed up to the Moon in four hours by demons summoned by his mother, an Icelandic witch: to survive the rapid motion he needed to be lean and fit, dosed with opiates, and have a damp sponge to breathe through. The inhabitants call the Moon Levania, and it seems to them the centre about which everything rotates. It has two very different sides: Subvolva, upon which the Earth shines, and Privolva, where it is never seen. The Privolvans endure long dark frosty nights of fourteen terrestrial days, and correspondingly hot days; they wander about in hordes and shelter in caves. Serpents abound. Subvolva is watery and much more temperate; its inhabitants see the enormous globe of Volva wax and wane; the craters are their fortified towns. Kepler worked out what their eclipses would look like, and how they would see the heavens. His readers were meant to pick up the idea that wherever we are in the universe we think we are at rest at the centre of things, and that the Earth is a planet and shines – as Galileo had inferred from seeing the whole globe of the Moon faintly lit up at new moon. Unfortunately, the first readers instead pressed the witchcraft charges against his mother.

Independently perhaps of Kepler, the well-connected Francis Godwin (1562–1633), bishop and son and son-in-law of bishops, antiquarian and nepotist, wrote in about 1611 a book also published posthumously, *The Man in the Moone* (1638). The hero is a Spaniard, Domingo Gonsalez, who trains geese to tow him in a flying chariot. To his alarm they head straight upwards, escaping from the Earth's gravity, and coast weightlessly for twelve days until they reach the Moon, the target of migrating birds, and home of spirits and locusts. After a time among the more serene lunar beings, he returns homesick to Earth. Godwin was inspired by Gilbert, and perhaps Robert Burton, whose classic *Anatomy of Melancholy* (1628) is a marvellous, learned compendium about our crazy world, at which, with Democritus, we should laugh. Like Kepler, Burton appended a tabular synopsis to help the reader navigate his discussion of writings on spirits, stars and planets, and Utopias, including *The City of the Sun* (*La citta del sole*) by della Porta's associate Tommaso Campanella (1568–1639), a Dominican from Calabria in disgrace because of his astrology, anti-Aristotelian empiricism, hatred of the Spanish rulers of Naples and defence of Galileo. Released from gaol by Urban VIII in 1626, he fled Rome in 1634 for Richelieu's France.

The young John Wilkins made his scientific debut in 1638, promoting the idea that the Moon was another world. He told his readers how science was a matter of probability, but his *Discovery of a World in the Moon* was not fiction like Godwin's, and he believed that it really might be possible to get to the Moon. Just as magnets are only effective at short range, so he thought that the Earth's gravitational pull would cease about 20 miles (32 km) up: if future generations could get men that high, all would be plain sailing. In a sequel arguing that the Earth is a planet, the future bishop warned readers that the Bible, teaching religion and morality, took for granted the science (or common sense) of its day:

> The Holy Ghost in many other places of scripture, does accommodate his expressions unto the error of our conceits: and does not speak of divers things as they are in themselves, but as they appear unto us ... divers men have fallen into great absurdities, when they have looked for the grounds of philosophy from the words of scripture.[3]

This doctrine of accommodation was vital to the Scientific Revolution. It was not new, but adopting it was easier now because it chimed so well with the early-modern belief that they knew more than their ancestors.

In principle, *infinite* was an adjective properly applied only to God, and *indefinite* was applied elsewhere, but Henry More (1614–87), Cambridge Platonist and important influence on the young Newton, wrote in his poem *Democritus Platonissans* (1647) a passage on the Infinity of Worlds:

> I will not say our world is infinite,
> But that infinity of worlds there be;
> The Centre of our world's the lively light

Of the warm sunne, the visible Deity
Of this externall Temple.[4]

Other suns are centres for their own planets too. Wilkins had used the
phrase 'plurality of worlds', which, while avoiding the problem of infinity,
implies that we are not alone in the universe. This was one of the many
heresies for which Giordano Bruno (1548–1600) had been burned by the
Inquisition, and on which Galileo had prudently been silent. If there really
were rational beings on other planets, would they have fallen into sin? If so,
would Jesus' death have redeemed them, or would they need Saviours too?
Wilkins was safe from the Inquisition, and so in France were two literary
men: the raffish Cyrano de Bergerac (1619–55), who read Godwin in
French translation in 1648 and published his own highly entertaining book
about the Moon-dwellers, who eat by smelling, pay in poetry, bottle light
and talk in music; and Bernard Le Bovier de Fontenelle (1657–1757),
whose *Plurality of Worlds* (*Entretiens sur la pluralité des mondes*, 1686)
propelled him into the position of permanent secretary of the Académie
Royale des Sciences – since he lived into his hundredth year, the title was
appropriate. In flirtatious conversations between a marquise and a savant,
he popularised the Copernican system in Cartesian form. Since climate
and character went together, his Venusians were 'little black People,
scorched with the Sun, witty, full of Fire, very amorous, much inclined to
Musick and Poetry, and ever inventing Masques and Tournaments in
honour of their Mistresses'.[5] Fiery Mercury was the very Bedlam of the
universe, while chilly Saturnians were phlegmatic. Fontenelle made
astronomy delightfully entertaining, conveying the idea that the world was
a great machine, and all the more wonderful for it.

The first president of the Académie was Christiaan Huygens (1629–
93), a pioneer of the wave theory of light, who observed Saturn's rings
through his improved telescope, greatly enhanced the accuracy of watches
and applied to clocks Galileo's discovery, supposedly made while watching
the gently moving chandeliers in Pisa cathedral, that pendulums keep time
even if some are swinging higher than others. Swing-swangs became scientific toys, and grandfather clocks scientific instruments, in consequence. A
Protestant, Huygens prudently returned to the Netherlands in 1681 as

Louis XIV prepared to abandon the toleration that had prevailed since the Edict of Nantes of 1598, and invaded the Low Countries: Huygens' father and brother held the important office of secretary to the princes of Orange, and were leading figures in the Dutch Renaissance. His *Cosmotheoros* was published posthumously, and translated into English, in 1698. This final work by a great man was to be no 'pretty Fairy story' but set out serious thoughts about planetary inhabitants and the uniformity of nature. Copernicus' system is 'commonly receiv'd … and very agreeable to that frugal simplicity Nature shows in all her Works': anyone who denies it either fails to understand it, or 'has his Faith at another Man's disposal, and for fear of Galileo's fate dare not own it'. The Earth is a planet, and the other planets behave so similarly that most probably they are as beautiful and well stocked as Earth. The Creator would provide for animals there too; they might be very different from ours, but Nature's variety is limited: there must be water (adjusted in properties to the climate) to nourish the plants and animals, and some spectators to enjoy them, beings endowed with reason, with virtues and vices like ours, taking pleasure in eating and in sex, enjoying social life and music, equipped with hands, living in houses, and having geometry and the arts that flow from it. If not, the planets and their moons would be useless. Inhabitants of the Sun would have to be inconceivably different; but the fixed stars are suns, and will have their inhabited planets too: 'So many Suns, so many Earths, and every one of them stock'd with so many Herbs, Trees and Animals.'[6] In truth, Huygens' universe seems a little dull after those of Godwin, Cyrano and Fontenelle; but all these books stimulated debates, sometimes heated, about our place in the world that continue to our day.

Huygens taught mathematics to Gottfried Wilhelm Leibniz (1646–1716) and, when visiting England to lecture on gravitation at the Royal Society in 1687, met Newton, whose *Principia* appeared in that year. Ever since the Babylonians, the physics of the Earth and the heavens had been distinct. Moreover, the explanation of why planets move around, and the calculations that predicted events like eclipses, had been at odds. Ptolemy's epicycles and other devices permitting accurate forecasts conflicted in spirit with the principle of perfect circular motion; Galileo's circular inertia was incompatible with Kepler's elliptical orbits; and the same was true for

Descartes' vortices, to which Huygens and most French savants adhered. In the plague year of 1665 the students at Cambridge were sent home, and Newton got on to the differential and integral calculus, the nature of light and colours, and the theory of gravity: the major topics for which he is remembered and which, along with theology and alchemical chymistry, were to preoccupy him for the rest of his long life.

Newton, back in Cambridge, had much to work out and needed more data; and it was his optical work that was first published. He demonstrated in a 'crucial experiment' that a prism does not impart colour to white light: rather, it splits it into its component colours, so that a second prism will merely bend light of any one colour falling on it. Because different colours are thus refracted by different amounts, telescopes that depend on refraction would, he believed, inevitably produce fuzzy images surrounded by coloured fringes; and he invented and made a reflecting telescope with a curved mirror. He sent his paper, and the telescope, to the Royal Society, where Hooke (who was responsible for experiments and peer review) criticised it. Newton was awkward, brooding and reserved, saw his work as above controversy, and declared that he would keep his thoughts to himself in future. Nevertheless, in 1684 Hooke, Wren and the younger astronomer Edmond Halley (1656–1742) met in a coffee house and discussed the force that kept the Moon and planets in orbit. Hooke (who had earlier written to Newton about it) said that it varied inversely as the square of the distance between the Earth and the Moon, or the Sun and a planet. He couldn't prove it, and Wren offered a book worth forty shillings to anyone who could. Halley visited Cambridge, where Newton told him he had proved it years before: Halley persuaded him to return to dynamics, and arranged for the results to be published through the Royal Society. Because the Society had exhausted its funds, Halley (whose father was a wealthy soapmaker, prominent in the City of London) had to provide the financial guarantee demanded by the publisher, as well as tactfully coaxing the parturient Newton. Communicated to the Royal Society in 1686, the book appeared in 1687 with the imprimatur of the secretary to the navy and diarist Samuel Pepys (1633–1703) in his capacity as president – for censorship was enforced in Britain from King Charles II's Restoration in 1660 until the Revolution of 1688–9 and the Society was one of the bodies licensed to authorise publication.

A substantial quarto, over five hundred pages in Latin, preceded by a Latin ode by Halley, it was full of theorems and demonstrations in Euclidean form, and there were few who could follow it all. There are three sections, or books: the first formally sets up an inertial physics, and includes 'scholia' on time and space; for Newton, there was vulgar time, based on sundials, but for science there was absolute, true and mathematical time that flowed equably. Similarly, there was absolute space, a Cartesian grid extending to infinity in all directions, in which the Earth really (not just relatively) moved. Within this framework he defined his terms, set out his laws of motion and deduced the consequences: we call this Newtonian mechanics. The second book does what Descartes failed to achieve, examining and testing the idea that space is full of swirling vortices that carry the planets. This involved pioneering work in hydrodynamics, leading to the conclusion that such a system would entail an æther as dense as the planet, and was anyhow incompatible with Kepler's three laws and thus with what is observed: the hypothesis, as he put it, was pressed with difficulties. In the third book, he announced his law of gravitation, and applied it and the laws of mechanics to the world.

The elderly Newton told William Stukeley (1687–1765), freemason, antiquary, physician and clergyman, about the apple (the fruit of the Tree of Knowledge in the Garden of Eden) that gave him the idea of gravitation. That story is therefore better founded than that of Galileo dropping things from the Tower of Pisa, but both function as myths: weights, apple and Moon are subject to the same force, and behave alike – they fall. But whereas the apple hits the ground, the Moon is further away, where the force is attenuated, and is moving rapidly. If the Earth exerted no force upon it, it would go off in a straight line and be lost to sight; but its inertial motion combined with the Earth's pull keeps it in orbit. Newton showed that an ellipse (of which a circle is a special case) is the consequence of such motion, with the Earth or the Sun at one focus. For planets, because gravity is universal, it is not only the Sun that attracts them: they pull one another too. Kepler, using Tycho's data, was lucky in that the observations were too accurate to be fitted to circles, but not so accurate that they showed all the wobbles (detected with telescopes) that result from this mutual attraction, in accordance with Newton's theory. Indeed, in the nineteenth century the existence of the planet Neptune was inferred from otherwise unexplained wobbles in the orbit of Uranus. Armed with the theory, Newton and Halley were able to show that comets travel in exceedingly elongated ellipses: Halley predicted the return of the one now named after him, knowing he would not be alive to see it. This was a further blow to the notion of vortices. Newton put at the beginning of book three, 'Mundi Systemate', a list of 'hypotheses' that includes Occam's razor, the uniformity of nature and Kepler's laws. These assumptions were the basis upon which the phenomena were to be explained. In later editions of the book, Newton famously wrote 'hypotheses non fingo' – I do not feign hypotheses. By then the latter term had come to imply guesswork, and the previous list was replaced by 'Rules of Reasoning' and 'Phenomena'. How bodies could attract each other across void space remained a great mystery. Nevertheless, Newton's physics did not seem to his disciples a matter of probabilities, but of certainty. For the Scottish poet James Thomson (1700–48), God was a geometer, and Newton had read his mind:

> O unprofuse magnificence divine!
> O wisdom truly perfect! Thus to call
> From a few causes such a scheme of things,
> Effects so various, beautiful and great,
> An universe complete! And O beloved
> Of Heaven! Whose well purged penetrating eye
> The mystic veil transpiercing, inly scanned
> The rising, moving, wide-established frame. . . .
> The heavens are all his own, from the wide rule
> Of whirling vortices and circling spheres
> To their first great simplicity restored.[7]

Moreover, Newton's universe had to be infinite, otherwise stars at the edge would be attracted by the gravitational force of those further in and everything would collapse into one vast mass.

Newton's data came from the observatories set up at Greenwich, with John Flamsteed (1646–1719) as the first astronomer royal, and in Paris under Jean-Dominique Cassini (1625–1712), founder of a dynasty of

astronomers. Louis XIV and Charles II founded them to improve navigation; in particular, to help sailors find longitude. The meticulous Flamsteed,
who had had to buy his own telescope, would not release his observations
until he was sure of them, making Newton so impatient that he and Halley
seized and published data from Greenwich when Flamsteed was away,
provoking a major quarrel.

Halley had made his reputation with a voyage to St Helena, a stopping
place for East India Company ships, to observe the southern stars; later he
made two voyages in command of HMS *Paramour* to map compass variation in the South Atlantic. The hope was that the compass deviation from
true north, thought to be due to the pull of continents, might be a clue to
longitude. The hope turned out to be vain, and one voyage ended in mutiny:
the Royal Navy decided that the days of gentleman captains commanding
'tarpaulin' sailors were past, and that when men of science travelled on naval
vessels in future they would be passengers under the command of a seaman.
For taking bearings and heights of stars, John Hadley (1682–1744), who
had in 1721 made an improved reflecting telescope in what became the
standard design, invented in about 1730 a reflecting quadrant (strictly an
octant) in which two images were made to coincide and the angle between
them was read off a scale. This was much more comfortable to use and
accurate than eye-straining earlier instruments.

Accuracy in engraving scales was improved by using dividing engines
rather than hand and eye; and in reading them by the introduction from
1631 of the device invented by Pierre Vernier (1580–1637) in which two
engraved lines are aligned. But wobbling remained a problem, and in observatories clocks, by makers such as Thomas Tompion (1639–1713), regulated as Huygens recommended by a pendulum, proved to be the answer
– though, in contrast to watches, regulated by a balance but inaccurate, they
could not be carried around or taken on board ship, and were no help to
navigators at sea. In observatories, however, the telescope could be fixed,
and the exact time when the star crossed its crosswires recorded instead of
an angle. Clock- and instrument-making became an important trade in the
eighteenth century, as observatories proliferated, surveyors mapped with
more precision, navigators charted havens, teachers demonstrated natural
philosophy and virtuosi kept records of the weather.

The naked eye is a wonderful instrument, dilated upon by natural theologians; but in the Scientific Revolution telescopes (based upon magnifying glasses) brought the distant scene close, and in a parallel development microscopes revealed an invisible world. Science ever since has been based upon the senses augmented by apparatus. In 1664 Henry Power (1623–68) published *Experimental Philosophy*, of which the most striking part was its microscopic observations, stimulating Pepys to buy a microscope for his own use. But the book was almost unillustrated, and it was Hooke's *Micrographia* (1665), with its magnificent plates, that called the attention of the world to the wonders that had formerly been too small to see. With scientific detachment, he carefully observed a gnat biting him before portraying her through his instrument. Hooke's enthusiastic language, the illustrations in which he made good use of his knowledge of portrait engraving, and his digressions into natural philosophy and corpuscular theory, made the book a classic, and soon natural history was being transformed by investigations of the life-cycle of insects and the sex of plants.

The most powerful microscopes of the seventeenth century were made by Anton van Leeuwenhoek of Delft, who made very small but powerful single lenses which he fixed into a brass plate. With them, he observed spermatozoa, and then bacteria and protozoa; but mammalian ova were not identified until the nineteenth century, and there was no reason (since his samples came from healthy people) to connect bacteria and disease. Microscopes were not easy to use: getting sufficient illumination was a problem, and the images produced were fuzzy, making it even trickier than with telescopes to distinguish what was really there and what was merely the effect of technical imperfection and human imagination. Science is often being done at the limits of accuracy.

Newton's world is sometimes called the clockwork universe, but mechanical philosophers in the tradition of Descartes were disappointed with him. As Leibniz put it to Newton's disciple Samuel Clarke (1675–1729), space, time and motion were not absolute but relative: 'if there were no creatures, there would be neither time nor place.' Moreover, action at a distance was inexplicable: 'He will have the sun to attract the globe of the earth through empty space? Is it God himself that performs it? But this would be a miracle, if ever there was any.'[8] The retort – that gravitational

attraction was a fact of nature, and that miracles have to be one-off –
seemed to miss the metaphysical point: space had to be void to allow the
endless circling of planets and comets, but how could one lump of matter
act upon another if there was nothing between them? Newton himself was
puzzled about it, and raised theological issues that we shall come to later:
but as a long-time adept in alchemy, aware of the affinities and attractions
displayed in chymistry, he was not a convinced mechanist like Descartes or
most contemporary natural philosophers. Because Newtonian physics
worked, the objection that its basis was 'occult' had little force, especially in
Britain and the Netherlands where he was hugely admired. In the preface
to the second edition of the *Principia* (1713), Roger Cotes (1682–1716),
Newton's disciple and editor, mocked the mechanists:

> *Galileo* has shown that . . . a stone projected moves in a parabola. . . . But
> now somebody, more cunning than he, may come to explain the
> cause. . . . He will suppose a certain subtile matter, not discernable by
> our sight, our touch, or any other of our senses . . . carried with different
> directions . . . describing parabolic curves. . . . The stone, says he, floats
> in this subtile fluid, and following its motion, can't choose but describe
> the same figure . . . should we not smile to see this new *Galileo* taking so
> much mathematical pains to introduce occult qualities into philosophy,
> from whence they have been so happily excluded? But I am ashamed to
> dwell so long upon trifles.[9]

If mathematics was indeed the language of nature, then those demanding
explanation in other terms could be regarded as foolish and ignorant: in the
eighteenth century, they were the losers as the witty Voltaire (1694–1778)
played a major part in bringing the French into line. 'Metaphysics' became
a term of abuse. Much later, with field theory, quantum physics, relativity
and cosmology, the argument began again.

By the time Newton died, in 1727, nobody seriously doubted any more
that the Earth went round the Sun; and Newtonian mechanics guides our
ordinary life and indeed our 'rocket science' today, despite quantum theory
and relativity. With the assistance of new instruments, the Atlantic became
a sphere of prosperity to rival the Indian Ocean. Its loathsome triangular

slave and sugar trade was a major component, but so were ships full of trade goods and emigrants going to and from the American colonies of Spain, Portugal, Britain, France and the Netherlands. Science was not only the key to understanding the world, it was also beginning to transform it. But it could still seem absurd, as in *Gulliver's Travels*, published under a pseudonym by Jonathan Swift (1667–1745) in 1726, where the island of Laputa is governed by dotty men of science. And while astronomy in some ways made the running in the Scientific Revolution, it was by no means all of it: we must now turn to the venerable science of chemistry.

CHAPTER 5

Interrogating Nature
The Use of Experiment

As a little boy Newton had done painful experiments, pushing a
bodkin into his own eyes while investigating vision, and had made
kites, toys and clockwork. He grew into the man who masterfully elucidated light and colours with his prism, but who also spent many hours
with furnaces and crucibles in alchemical labours. Like Galileo, but unlike
many theoreticians, Newton was a great experimentalist. In his laboratory he was confronted with a universe full of mysterious powers,
described in the kind of esoteric and arcane language that he loved to
wrestle with in the prophetic books of the Bible, but also with long
traditions of crafts or 'mysteries' and of practical experience with ores and
metals, and with pigments, perfumes, cosmetics and condiments made
from animals, vegetables and minerals. There were two different kinds of
experiment involved, corresponding to travelling and exploring: the
crucial experiment to decide under controlled conditions between two
possible hypotheses, proving, for example, that the prism separates rather
than generates colours; and the open-ended investigation of phenomena,
trying things and seeing what happens, like chymists determining
the properties of a substance. Newton did both, and in the later editions
of his immensely influential *Opticks* (1730) we find him pursuing chains of
experimental reasoning and leaving tantalising loose ends, in the form of
'queries', for his successors:

Have not the small Particles of Bodies certain Powers, Virtues, or Forces, by which they act at a distance ... upon one another for producing a great Part of the Phænomena of Nature? For it's well known, that Bodies act upon one another by the Attractions of Gravity, Magnetism, and Electricity; and these instances shew the Tenor and course of Nature, and make it not improbable that there may be more attractive Powers than these. For Nature is very consistent and conformable to her self. How these Attractions may be perform'd, I do not here consider.[1]

He wrestled both with the abstract problems of the running-down of the universe, and attraction at the cosmic and atomic levels, and with Huygens' rival wave theory of light (which he thought could not explain why it went in straight lines) as he struggled to account for puzzling phenomena that experiments revealed. It seemed to many that he had said the last word about mechanics, but had opened the door to a new world of experimental physics, an ocean of undiscovered truth for successors to explore.

Physics was not the first experimental science, for chemistry was already very old. Cooking, making pots and extracting metals were great achievements of our remote ancestors, requiring mastery and management of fire and matter. Chemistry is always hands-on, and involves all the senses: it is the science of Galileo's secondary qualities – colours, smells, tastes, textures and even noises. By 1500 mining for ores (and increasingly for coal) was a major industry in Europe, notably in the Harz Mountains and in the north of England, in Cornwall and in Spain, and soon to be carried on across the Atlantic in Spanish Mexico and Peru. The labours of chymists were far more relevant to and significant for the mass of mankind than knowledge of planetary orbits. Optimistic alchemists persisted in their labours despite failures, and perfected techniques of distillation, dissolution and recrystallisation. Antimony, in its 'mercurial' metallic or 'sulphurous' grey powder forms, was particularly interesting, and could be obtained in stunning star crystals: surely a step on the way. However sure we may feel that they cannot have generated gold, some thought that they had, and convinced reputable bystanders that they had witnessed a transmutation. Making alchemical gold was tempting but widely illegal (in England, for example, from 1403) because of fears for the monetary system and for honest goldsmiths; Boyle was later instrumental in getting this early legislative brake on research repealed.

Alchemy had been mocked as a realm of sleight of hand and chicanery by Chaucer, and then in 1610 by Ben Jonson in his play *The Alchemist*, a comedy about a pair of confidence tricksters. The prominent physician and chemist Herman Boerhaave (1668–1738) of Leiden is often credited with discrediting and dismissing alchemy; but Peter Woulfe (1727–1803) from Ireland was both a Fellow of the Royal Society and well known as an alchemist in the old tradition – and as exemplars he had Newton and Boyle. He stuck prayers written on slips of paper to his apparatus as talismans to ensure experimental success: hard to achieve when (before pyrex) glassware had to be both thin enough to stand up to heating, and thick enough to be reasonably robust to handle. It is not possible to falsify a high-level theory like alchemy (or 'creationism') experimentally, but it can be made increasingly implausible, irrelevant and open to satire by poets and playwrights.

Paracelsus had established a tradition of chemical philosophy in which the Creation was seen as a chemical process, the chemist as uniquely qualified to interpret Genesis, and chemistry as the queen of the sciences, understanding and emulating Creation through experiment. Alchemists made much of male and female imagery in explaining their science, and were chary of barren reason, seeking mystical illumination through experience as an adept in the laboratory. They deeply revered the symbol of the uroboros, a snake biting its tail – a powerful image that can serve as a reminder of the cyclic processes so crucial in chemistry and life, but that also stimulated August Kekulé (1829–96), so he said, in thinking of a ring structure for benzene. Paracelsus sought to recover true science going back to Adam, with healing as its goal; he was a Luther, purifying science from pagan and scholastic corruptions, and restoring 'Mosaical Philosophy'. This was alarming to orthodox churchmen, but attractive to his disciples, who introduced Neoplatonic ideas into the mix of the spiritual and the material that underlay their chymistry. Religion and science were thus intimately engaged not only in astronomy, demonstrating the order in the world, and natural history, demonstrating the world's diversity, but also in chymistry. But in the sixteenth century chymistry remained a mystery, a craft, its processes described in language and symbols that only the initiated could understand: it was occult, not public, knowledge.

Nevertheless, from 1602, a great number of alchemical texts previously available only in manuscript were published in Latin as *Theatrum Chemicum*, with the best edition appearing in 1659–60 (six volumes, Strasbourg). Though usually obscure, these texts were now available for comparison and study: chymistry was beginning to come into the open, and this broadened its appeal beyond those with practical interests in metals or medicine. In the Neoplatonic tradition, Dee and the Paracelsian doctor Robert Fludd (1574–1637) delighted in the analogy between the universe or macrocosm, and mankind the microcosm. Elias Ashmole (1617–92), Royalist, astrologer, Fellow of the Royal Society and founder of the museum in Oxford that still bears his name, shared this world-view, and in 1651 published *Theatrum Chemicum Britannicum*, a collection of alchemical writings in English prose and verse, handsomely illustrated. Anyone seeking the wisdom and experience embedded in alchemy's long history now had readily accessible materials at hand in Latin and vernacular languages, even if they were hard to interpret.

To anyone working in this tradition, the most important chymist of the early seventeenth century was the Fleming Joan-Baptista van Helmont (1577–1644). A physician, and follower of Paracelsus, he found that when someone was wounded it was efficacious to treat the weapon with the appropriate unguents, rather than the patient, whose wound was simply cleaned and bandaged. To us, this shows something about the contemporary pharmacopoeia; but, to Helmont, the success of this 'weapon salve' established that in his pantheistic world where matter was ensouled and everything interconnected, the weapon was the cause of cosmic disturbance and thus required treatment. These ideas aroused the interest of the Inquisition, for Flanders was under Spanish Habsburg rule, and from 1625 Helmont was confined under house arrest – just like Galileo in contemporary Florence, but less famously so.

His influential *Ortus Medicinae* was therefore left to be published posthumously by his son Franciscus Mercurius in 1648, with an English translation following in 1662. Rejecting Aristotle's four elements, and impatient too with the sulphur, salt and mercury 'tria prima' of Arab alchemists and Paracelsus, he saw water as the fundamental substance: the notion goes back to the ancient Greek sage Thales of Miletus (c. 624–c. 545 BC), and accorded with Helmont's reading of Genesis 1. But Helmont tested the idea experimentally, taking a young willow tree and planting it in a tub with 200 pounds (91 kg) of earth. He nourished the tree with distilled water, protected it from dust, and collected and weighed the leaves that came off each autumn. After five years he uprooted the tree and dried the earth, which weighed the same as at the start of the experiment. From the water alone, the tree had grown from 5 to 169 pounds (2.3 to 76.7 kg) and had produced copious leaves. These could be burned, releasing 'fire' and leaving 'earth', revealed in the form of ash. So earth was not a primordial element; nor was the potent immaterial agent fire. For Helmont, air did not have a chemical role either, but he did identify as 'gas', from a Greek root meaning chaos, the effluviae from numerous chemical processes: 'the spirit, hitherto unknown, which can neither be retained in vessels nor reduced to a visible body'.[2] Gas, derived from burning charcoal, from seashells dissolved in vinegar, from fermentations and putrefactions, found in mines and caves, was distinct from air, which was composed ultimately of water.

Helmont also experimented with the 'water glass' that was used for preserving eggs during the Second World War. On fusing sand with alkali, he got a solid product which liquefied on standing in air and formed a sticky suspension in water; when he added acid, he got back the same weight of sand. He was fascinated by the process in which liquid turned reversibly into solid; here as with the willow tree, his experiment was quantitative. Practical men had had to take note of costs and yields as they refined metals and manufactured ceramics or dyes, and artillerymen worked out the best proportions of sulphur, saltpetre and charcoal for gunpowder. But businessmen and soldiers have always had to keep trade secrets, and alchemists had been both secretive and usually qualitative. Helmont's experiments could be repeated (if not exactly) by others: chymistry was becoming definite and public knowledge. In chemistry, one of the great problems has always been distinguishing what is crucial from what are side-reactions due to impurities or particular conditions, akin to what in another context are called 'message' and 'noise'. Problems with variable and uncertain reagents make it difficult for us to repeat long-past experiments:

chemists since about 1900 have become used to having pure samples handily in bottles, whereas our ancestors had to refine and purify their raw materials as best they could. Boyle's descriptions are full and clear enough to be 'translated' and yield the colours, tastes and smells he reports; Helmont's might be too.

Helmont's view that water was the ultimate element was not uncontested, even by those who shared much of his unmechanical world-view. Cornelius Agrippa (1486–1535), who was simultaneously a philosopher in the sceptical tradition and an occultist, had seen earth as fundamental. Coming from Cologne, he went to Italy in 1511 where he encountered a rich mixture of Florentine Neoplatonism, Jewish Cabbala and the Hermetic corpus; thence to France, and on to Antwerp, where his books on the vanity and uncertainty of knowledge and on occult philosophy were published. Although Franciscan and Dominican critics did their best to demolish them, his unorthodox ideas proved influential. The Welsh metaphysical poet Henry Vaughan's twin brother, Thomas (1621–66) – Rosicrucian, Royalist and Oxford graduate – admired Agrippa's writings and became an alchemist, taking the name Eugenius Philalethes.

Whatever one thought about such speculations, Helmont's experimental methods made him an excellent role model. He brought into chymistry the precision that had been necessary in pharmacy and industry, and especially for assayers and goldsmiths, whose livelihoods depended upon accurate weighing and measuring. And experimenting might bear fruit. Faced with the ruin of the German lands during the Thirty Years' War, Johann Rudolph Glauber (1604–70) hoped that chymistry would not only restore but increase their prosperity. The *Works* of this 'highly experienced and famous chymist' were translated into English 'for publick good' in 1689. Containing 'choice secrets' in medicine and alchemy, mining and metallurgy, the making of saltpetre, the improvement of barren lands and crops, the book would be 'very profitable for lovers of art and industry'.[3] Glauber, who worked in pharmacy and mirror-making, is remembered chiefly for 'his' salt, hydrated sodium sulphate: specific words like 'salt', 'sugar' and 'alcohol' have become, across the history of chemistry generic, applied to a wide range of similar substances. Glauber was an important figure in promoting chymistry, which, indeed, transformed into modern chemistry in the eighteenth century, played a crucial role in securing the nineteenth-century prosperity of Germany.

Helmont's world-view was widely shared in the first half of the seventeenth century, notably among religious radicals. But when he died, people (including some of those who shared his world-view) were also beginning to take atomic theory seriously. The manuscript of Lucretius' great poem *De Rerum Natura* came to light in Florence in 1417, and was printed and published in Brescia in about 1470. It was greatly admired for its style and language, but seemed absurd or horrifying in its atheism, materialism and basis in blind chance. For Lucretius, indestructible atoms had always existed, and they interacted by hooking together and separating in unpredictable collisions without purpose or end. In Antiquity, Aristotle had denounced Democritus' earlier version of atomism because it failed to make any sense of the manifest order of things in the living world, notably in embryology; Platonists too believed in a law-governed cosmos, governed by Reason. The same factors made atomism unacceptable in the early-modern world: the Scientific Revolution was based upon finding law, not chance. The text remained exemplary, but its message seemed preposterous – and yet fascinating, toyed with by Galileo and Harriot, and taken up by the sceptical essayist Montaigne among others.

It was Pierre Gassendi (1592–1655) who made atomism respectable, indeed central, in science. He was a Roman Catholic priest, based in Provence but frequently to be found in Paris. A diligent astronomer, he observed the transit of Mercury across the Sun in 1631: he was himself a Copernican, though admitting that it was hypothetical. His atoms were created by God, who could split or annihilate them, and who determined their properties and behaviour; but for human purposes, they were permanent. These atoms of Lucretius and Gassendi were not like the atoms of modern chemistry – hydrogen, helium, lithium and so on through our Periodic Table – all irreducibly different: they were fundamental particles, all made of the same stuff, and in different combinations and arrangements made up the very stable metals, non-metals and then all the other things that compose the world. Like Descartes, Gassendi was a mechanist, but his beliefs about matter were very different: instead of a world of matter, a plenum, he had matter and void.

Atomism was readily compatible with alchemy, for transmutation could result from the deep rearrangement of particles. George Starkey (1628–65), brought up in Bermuda and New England, went up to Harvard in 1643, graduating AB (Bachelor of Arts) in 1646 and subsequently AM (Master of Arts), and then practising medicine in Boston, was such an alchemical atomist. In 1650 he came to London, already with a reputation as a savant and practitioner; he made perfumes and pharmaceuticals, chymical furnaces and other equipment, joined the circle of improvers and projectors around Hartlib, and established a medical practice. In January 1651 he met Robert Boyle and introduced him to chymistry, inspiring in him a devotion to empirical laboratory science and bringing him up to date with Helmontian ideas. Boyle's lawyer father, Richard (1566–1643), had gone to Ireland in 1588 when land confiscated during a rebellion was being redistributed among English colonists; he married heiresses, bought Sir Walter Raleigh's Irish lands and prospered exceedingly. He promoted English settlement in Ireland, held high office, was created earl of Cork and became rich; although he was the youngest son, Robert Boyle was comfortably off, managed to keep out of the Civil War and might have been expected to be a patron rather than a distinguished practitioner in science. In contrast, his mentor and protégé Starkey overreached himself, and in 1654 was imprisoned for debt. Thereafter he re-established himself and was a notable figure in learned circles in the tumultuous years of Cromwell's Protectorate and Charles II's Restoration, dying in the Great Plague in 1665. Unlike many physicians, he did not desert his patients during the epidemic; but his remedies did not protect him from it. In his writings he adopted, as alchemists often did, a pseudonym, Eirenaeus Philalethes, indicating his peaceable love of truth. He was a follower and admirer of Helmont, and like him sought the alkahest, or universal solvent, as well as 'sophick mercury'; but, unlike Helmont, he incorporated atomism readily into his chymistry. His writings were read by Boyle, Newton and John Locke (1632–1704).

Gassendi's atomism came chiefly to English readers, however, through Walter Charleton (1619–1707) and his handsome quarto *Physiologia Epicuro-Gassendo-Charletoniana* of 1654. Wilkins had been Charleton's tutor at Oxford, where, turning to medicine, he received his MD at the age of twenty-two, and shortly afterwards became physician to Charles I, then in Oxford as the Civil War began. He picked up and translated Helmontian chymistry in his *Ternary of Paradoxes* (1650), and became a close friend of the remarkable Sir Kenelm Digby (1603–65), dashing courtier and diplomat, Paracelsian chymist and Catholic recusant. After the king's execution, Charleton was appointed physician to the exiled Charles II but remained in England, going in 1649 to London, where he wrote and practised. He was part of a circle with links to Marin Mersenne in Paris that included Thomas Hobbes (1588–1679) and the eminent physician George Ent (1604–89), a close friend and supporter of William Harvey, and, like him, a graduate of Cambridge and Padua. Charleton and Ent were founding Fellows of the Royal Society, and in 1689–91 Charleton was president of the Royal College of Physicians. He later held other offices in the College, but in old age his practice declined and he seems to have died in poverty. His most famous book, published in 1663, was on Stonehenge, which he (following contacts with the Danish antiquary Ole Worm, 1588–1654) attributed to the Danes; but *Physiologia*, in which he deserted Helmontian pantheism in favour of Gassendi's mechanistic atomism, was his most influential.

This book is an argument, sometimes pitting 'Atomist' against 'Anti-atomist', elsewhere set out in numbered sections or 'articles'. Charleton's atoms differed in shape and size, like the letters of the alphabet:

> For as *Letters* are the *Elements of Writing*, and from them arise by grada-tion, Syllables, Words, Sentences, Orations, Books: so proportionately are *Atoms* the *Elements of Things*, and from them arise by gradation, most exile [smallest] Moleculæ, or the Seminaries of Concretions, then greater and greater Masses successively, until we arrive at the highest round in the scale of Magnitude.[4]

He was fascinated by crystalline forms, which might indicate these atomic shapes; in classical fashion, he postulated atomic hooks, as indestructible as the atoms themselves, rather than mysterious attractive forces or immaterial agencies. Matter and motion, and action by contact, underlay his uncompro-misingly mechanical world. He was confident it would replace the uncertain and contentious views of the quarrelsome philosophical sects as a firmly founded basis for science. This was a view that Boyle was to champion.

John Dalton (1768–1844) is celebrated because his later version of the atomic theory in chemistry had a quantitative experimental basis and consequences. In fact, the arguments of Boyle and his predecessors for atoms, or corpuscles as Boyle preferred to call them, had already convinced most; but such atomism could explain only in principle, not in detail, and Antoine Laurent Lavoisier (1743–94) in 1789 condemned it as metaphysical. Boyle's best-known book is *The Sceptical Chymist* (1661), an indigestible and pedestrian dialogue reprinted in Everyman's Library in 1911, although by the twentieth century it hardly made for easy reading. Boyle was concerned to show how neither the Aristotelian elements, nor Paracelsian principles, nor Helmont's basis in water account for the facts of chemistry. Aligning himself with Francis Bacon, he initiated a long tradition of sceptical scientific empiricism, careful exploring rather than rash travelling, following Helmont's quantitative experimental lead.

That can be seen in his investigations of heat and cold, gems and crystal forms, and colours – about which he wrote, characteristically: 'If I sometimes seem to insist long upon the circumstances of a Tryall, I hope I shall be easily excused by those that both know, how nice [tricky] divers experiments of Colours are, and consider, that I was not merely to *relate* them, but so as to teach young Gentlemen to make them.'[5] The scrupulous experimentalist, keen to recruit fellow explorers, was also anxious to promote intelligible explanations of the phenomena he observed, and in *The Origine of Forms and Qualities* (1666) he argued more clearly than in *The Sceptical Chymist* for matter and motion as the basis of science. Atomic theory was important because only mechanical explanation would ultimately satisfy the scientific mind; for Boyle, it posed no threat to his strong religious faith. He admired clockwork, wonderful mechanisms like a little world, and he strove to bring chymistry within the respectable fold of mechanical philosophy, for which he became the leading British exponent. His piety removed the taint of materialism from atomism.

Boyle did not, however, carry all his contemporaries with him. For Samuel Duclos (1598–1685), chymist and founder member of the Parisian Académie des Sciences, this mechanical programme did not elevate chymistry to the ranks of real science, but was on the contrary reductive and useless, not really explaining chymical phenomena. He charged Boyle with being somewhat ignorant and incompetent as a chemist, not respectful enough to understand the authorities and develop the tradition in a discipline properly distinct from physics. For the same reasons, he took on the president of the Académie, the ardent mechanist Huygens, claiming that he could not account for the difference between congelation, as when water freezes to ice, and coagulation, as when cheese is formed from milk left to stand. For Duclos, they were irreducibly different: freezing is a physical and more superficial process than coagulation, a chemical process that eludes corpuscular explanation. Argument, debate and demarcation disputes are as characteristic of science as is its progress. Indeed, they are important features of it, and revolutionary proposals, even when made by eminent practitioners, never convince all contemporaries.

Boyle sought to resolve controversy and extend knowledge through experiment. It was said of Davy that his greatest discovery was Faraday, and of Boyle we could say that it was Hooke. A clergyman's son, on an Oxford scholarship as a chorister at Christchurch, he was a mechanical genius of immensely wide interests who made many of Boyle's experiments possible, and indeed carried many of them out. He is remembered for 'Hooke's Law', we have met him as a microscopist, and he was also a major geologist and architect. His reputation has suffered in part thanks to the dislike of Newton (who long outlived him – a good way of triumphing over enemies) and the fame of Wren, who eclipsed his architectural achievements; he is neglected in the way the polymath (the pentathlon athlete, the all-rounder) is less remembered than the specialist, and owing to a preference among philosophical historians for the theorist over the practical experimentalist. In 1666 after the Great Fire, he collaborated with Wren, designing the Monument that stands in the City, various City churches (often attributed solely to Wren) and Bethlehem Hospital for the insane, notorious as Bedlam. Stories about his quarrelsomeness would appear to be projections from Newton; while he was lean and bent, he seems to have been convivial enough, though he was involved in acrimonious priority disputes, notably with Huygens.

Boyle was struck by chemical colour-changes and used cabbage juice to indicate acidity or alkalinity, an opposition that is very important in chemistry; but the experiments that caught the imagination of his contemporaries were done with the air pump that Hooke built for him. Galileo had been puzzled that suction pumps for water will not work from wells over 30 feet (9 m) deep; he thought the column of water broke under its own weight, but his disciple Evangelista Torricelli explained it otherwise. We live in an ocean of air that presses down upon us and forces water up into an evacuated space: if a long tube is filled with mercury and inverted, then because it is much denser than water the column will be only about 30 inches (75 cm) high. The mathematician and philosopher Blaise Pascal (1623–62) tested the idea in 1647 by getting his brother-in-law to take a mercury tube up a mountain, the Puy de Dôme, where he found that the mercury level fell as he climbed. The mercury tube was a barometer, measuring air pressure; it became a standard way of measuring heights, with scientific mountaineers lugging fragile glassware up to summits. Above the mercury was an empty space, which Torricelli and his colleagues in the Accademia del Cimento (Academy of Experiments) investigated. They demonstrated in a series of intricate experiments with different-shaped tubes (sometimes one inside another), leaning at different angles, that the column could not be held up by skyhooks, and argued that above it was a vacuum. If nature did not abhor a vacuum after all, the idea of atoms and void was less implausible.

Experiments with barometer tubes were awkward to perform, but in May 1654 Otto von Guericke (1602–86), engineer and burgomaster of Magdeburg, demonstrated before the Imperial Diet in Regensburg an air pump with which he evacuated a large hollow metal sphere made in two halves. Sixteen horses couldn't pull the sphere apart. His pump was inconvenient and slow, and Boyle resolved upon getting something better, with a transparent receiver so that effects could be observed. In 1659 Hooke duly made the pump that Boyle described in *New Experiments, Physico-Mechanical, Touching the Spring of Air and its Effects* (1660). A brass cylinder with a piston worked by a crank was surmounted by a glass globe with a valve at the bottom and a lid at the top. Boyle described more than forty experiments: in a barometer put into the receiver the level fell, a ringing bell could no longer be heard and fire was extinguished.

Demonstrating the air pump became a standard feature of the Royal Society's entertainment of visiting dignitaries; and air pumps (improved by Denis Papin, who added a second cylinder) became part of the equipment of itinerant lecturers, famously captured in the 1768 painting by Joseph Wright of Derby that shows a struggling bird in the receiver. To Boyle, by removing the air, the pump left a vacuum in the receiver; but he found himself having to argue against those who believed that it was full of æther: notably Hobbes, whose political theory set out in *Leviathan* (1651) had scandalised contemporaries, but also the English Jesuit Francis Line (1595–1675), whose 'funicular hypothesis' involved a skyhook holding up the mercury column. Hobbes' earlier claim that he had squared the circle drew him into a quarrel with the mathematician John Wallis (1616–1703). Hobbes was an armchair critic of science, but experiments (even Boyle's) do not strictly entail a theory, and to postulate void spaces stretched the seventeenth-century imagination. For Hobbes, as for Leibniz, philosophical coherence was much more important than experiment; and abhorrence of vacua went with Hobbes' advocacy of absolutism as the answer to anarchy and political chaos, curbing our ruthless selfishness.

Like Helmont, Boyle did not think of air as a chemical entity, but as a fluid in which particles essential in respiration or responsible for rusting would float. The fact that it was springy was particularly striking to him, and he investigated it further. Before the Royal Society, and then in the second edition of his *Spring of Air* (1662), Boyle displayed a 'Table of the Condensation of the Air', demonstrating what he called his hypothesis and we call Boyle's Law. He used a J-shaped glass tube with air trapped in the shorter leg (as he called it); by adding mercury to, or taking mercury from, the longer leg, the pressure could be varied. The tabulated results showed a close fit to 'the *Hypothesis*, that supposes the pressures and expansions to be in reciprocal proportion'.[6] In our terms, pressure times volume is constant. What is striking is that the figures calculated from observations tally closely but not exactly with the predicted ones: Boyle was presenting uncooked experimental data. Experiment and observation are always subject to error; laws based upon them can nevertheless seem certain. Boyle expected and wanted his work to be repeated, survive these tests and thus come to form part of an established body of scientific knowledge.

Pressure could be increased, decreased and measured, but our 'gas laws' – not only Boyle's Law but that of Jacques Charles (1746–1823) too – also involve temperature. Although Galileo had devised a way of telling how hot things were, his instrument had no definite fixed points and could not be exactly replicated; and it was not until the eighteenth century that temperatures could be unambiguously measured using thermometers with scales based upon the freezing and boiling points of water (which in practice turn out to be awkward to measure). The hymn has it that God guides the planets through laws that never shall be broken: in fact, we know that the 'laws of nature' that we learn in school are not definitive. There are lawbreakers, as carbon dioxide is for the gas laws – and as all gases are under extreme conditions; Kepler's laws of planetary motion are not exact, and Newton's laws (the rocket science that got astronauts to the Moon) also fail with speeds close to that of light. From the nineteenth century onwards, they have been shown to be idealisations that work for perfect gases, point masses and Galilean frames of reference to which the real world more or less approximates.

In 1675 Papin, who had worked in Paris with Huygens, came to England, and soon collaborated with Boyle in further experiments on the spring of air. An outcome of this was Papin's 'digester', the first pressure cooker, which was described in a short book of 1681, full of experiments and recipes, and carrying the recommendation that it would be particularly useful on sea voyages and for making 'the oldest and hardest Cow-Beef as tender and as savoury as young and choice meat'.[7] Because one cannot peep into a pressure cooker as one can into an ordinary pot, Papin devised ways of determining pressure and temperature (having no suitable thermometer): a safety valve with a weight that could slide along an arm, and a pendulum to time how long a drop of water took to evaporate. As well as cookery, the apparatus might, he hoped, be useful and economical for dyers extracting colours, and in perfumery. He was elected FRS (Fellow of the Royal Society) in 1680, but there was no career structure in science: seeking a job, he went to Venice in 1681, was back in England in 1684 but, unable to find a post, then went in 1687 to Kassel, where the landgrave of Hesse supported him and he devised a kind of steam engine, and finally came back to England in 1707 and died in obscurity. He could not return to his

native France, where there was a salaried Académie des Sciences, because
he was a Huguenot – a Protestant – and Henry IV's Edict of Nantes (1598),
which ended the wars of religion and brought toleration, was revoked by
Louis XIV in 1685, leading to massacres of Protestants and to their flight
and exile. Many were skilled, and this diaspora was comparable to that of a
hundred years earlier, when Antwerp surrendered to the Spanish army in
1585 and the science, trade, industry (notably printing), commerce and fine
arts of the Dutch Republic benefited enormously from the influx of fleeing
Protestants as the city lost 40 per cent of its population and its outlet to the
sea, blockaded for two hundred years.

Hooke, Boyle's greatest protégé, became the early Royal Society's resi-
dent experimentalist. His microscope had a tube handsomely bound round
in leather, which enterprising makers copied for up-to-date gentlemen to
have, like globes, in their libraries: they can be dated from their tooling
patterns. Hooke focused light with a globe of water and a lens onto his
microscope slides so that he could see more clearly, and vividly reported the
wonders he saw – for example, the eyes of a fly composed of many 'pearls':

> Now, though there may be by each of these eye-pearls, a representation
> to the animal of a whole *Hemisphere* in the same manner as in a man's
> eye ... yet some very few points which liyng [*sic*] in, or neer, the optic
> *Axis* are distinctly discern'd: So there may be multitudes of Pictures
> made of an Object in the several Pearls, and yet but one, or some very
> few, that are distinct ... I think we need not doubt, but that there may
> be as much curiosity or contrivance and structure in every one of these
> Pearls, as in the eye of a Whale or Elephant.[8]

And he was moved with aesthetic delight when he contemplated a flea:

> The strength and beauty of this small creature, had it no other relation
> at all to man, would deserve a description ... the *Microscope* manifests it
> to be all over adorn'd with a curiously polish'd suit of *sable* Armour,
> neatly jointed, and beset with multitudes of sharp pins, shaped almost
> like Porcupine's Quills, or bright conical Steel-bodkins; the head is on
> either side beautify'd with a quick and round black eye, behind each of

which also appears a small cavity in which he seems to move too and fro
a certain thin film, beset with many small transparent hairs, which probably may be his ears.[9]

This ecstatic description was carefully keyed to the famous engraving.
And Hooke hoped that just as 'glasses' had improved our seeing, so mechanical inventions might promote hearing, smelling, tasting and touching.

We think of Hooke, as we do of Faraday, as a physicist rather than a
chemist; but both took into their researches the careful experimental
approach characteristic of the chemistry in which they had begun. Hooke's
enthusiasm for Baconian mechanical philosophy shines through his preface
to *Micrographia*, where he hopes for an improved chymistry leading to
flying machines, and perhaps the alkahest; and in the copious praise for the
Royal Society and its luminaries Boyle, Wilkins, Wren and his precursor in
microscopy, Henry Power. *Micrographia* reveals his interest in crystal structures, the way they might be generated by the close-packing of spherical
corpuscles, and the colours observed in thin plates of mica.

In 1704 Hooke's *Posthumous Works* appeared in another folio volume,
including his lectures on Light and Colours, Method for Improving
Natural Philosophy and Discourse on Earthquakes. On the subject of
optics, he differed from Newton in seeing light rays as pulses, and colours
as modifications of white light rather than components of it; on method, he
argued very cogently, and clearly from long experience, not only for experiment and observation, but especially for recording 'The Histories also of
such as are conversant about Mineral, Vegetable, and Animal Substances',
in a wonderful range of about two hundred crafts and trades which reads
like an elaborate list for census enumerators.[10] He also listed series of
queries and tips for making discoveries. In *Earthquakes*, assembled from
papers read to the Royal Society between 1663 and 1699, which has superb
plates especially of ammonites and other fossils, he sought to account natu-
ralistically for the petrifaction and location of what to him were the remains
of real organisms: we see his mind in action as he thinks the matter through.

Posthumous Works also contains Hooke's review of Ashmole's *Theatrum
Chemicum Britannicum* and a discussion of Dee's writings on spirits, which
he considered to be cryptographic, concealing their true meaning, for which

the biblical book of Enoch might be a key. In his opinion Dee was employed by Queen Elizabeth I on a secret mission and was sending back coded information about Emperor Rudolph II. His famous crystal ball

> was of a considerable bigness, and was placed upon a Pedestal, or Table, which he calls a Holy Table, which might contain the *Apparatus* to make Apparitions, when he had a mind to be seen in it, as likewise to produce Noises and Voices, if there were occasion. All of which might be done by Art.[11]

As a secret-service man, Dee would have been a conjuror in our sense of the word: a trickster. Unlike microscopes, crystal balls did not readily lend themselves to repeatable experiments or straightforward explanation, as required by mechanical philosophers. But codes and their cracking were of great importance in a time of religious and civil wars, providing interest and employment for mathematicians and linguists, and making the apocalyptic books of the Bible seem especially relevant and in need of deciphering. If the end of the world was near, identifying Antichrist and deciding whether 'Babylon' meant Rome became topics of vital, serious, sometimes enraged argument. Alchemical texts had the same character of imparting and withholding information; a veil had to be penetrated. Making sense of what one saw through early optical instruments (or, indeed, crystal balls) was not wholly different: fuzzy images needed interpretation, like X-rays or scans. And while Hooke made the microscope familiar in Britain, it was in the Netherlands, where the telescope had been invented, that lens-grinding had first become an important craft, with Baruch Spinoza (1632–77) among its practitioners.

Glass, in a range of vivid and subtle colours, had been an important feature of the great Gothic cathedrals; as the Scientific Revolution began, clear glass for windows (as well as for tableware, chymistry and pneumatics) was transforming architecture and daily life, not only for grandees but also for townspeople. Glass was both sufficiently commonplace and luxurious for taxes upon windows to be imposed in seventeenth-century England. It was difficult to get rid of every streak and bubble in glass, and to make large panes; but lenses good enough for reading glasses were being made in the

Netherlands early in the fifteenth century. Spectacles were imported into England, and some makers settled in London; in 1629 the latter separated themselves from the Brewers Company with which they had been associated and acquired a charter as a separate Company, giving them the right to control entry into the trade, and standards. Venice, where Galileo got his telescope made, was a much more important centre for glassmaking. It is striking that while Galileo's images of the Moon and its craters look convincing, they are impressionistic: depicting and interpreting what could be seen through lenses, especially with instruments using two of them, was not straightforward. Descartes, doing geometrical optics, realised that lenses whose faces are part of a sphere, and thus easiest to grind, do not focus all light rays to a single point; to get clarity, one must use only the central part of the lens, but that means restricting the light and hence the power. He knew that parabolic or hyperbolic, rather than spherical, lenses (and mirrors) would focus more clearly, and he succeeded, in the Netherlands, in getting some made.

They did not answer his needs, however, because the coloured fringes around the image were just as bad or worse. Newton argued that these resulted from light of different colours being refracted differently, and were thus inevitable with lenses: hence he invented his reflecting telescope. John Dollond (1707–61), son of a Huguenot refugee silkworker, picking up optical ideas from others and making them practicable and profitable, was in 1758 given the Royal Society's Copley Medal for his compound achromatic lens. Dollond fitted together two lenses, one made with glass containing lead and thus more refractive, so that their effects upon colours cancelled each other out. Such lenses made refracting telescopes, which could be much smaller than reflectors of similar power, practicable, but the little lenses of microscopes were much harder to correct, and it was not until about 1800 that achromatic microscopes began to be available.

It was easy for the satirist Samuel Butler (1613–80) to make fun of a savant at the Royal Society inspired by Kepler's *Dream* of warring Moon-dwellers:

> With that a great philosopher,
> Admired, and famous far and near,
> As one of singular invention,

But universal comprehension,
Applied one eye, and half a nose
Unto the optic engine close.

Quoth he, 'A stranger sight appears
Than e'er was seen in all the spheres,
A wonder more unparalleled,
Than ever mortal tube beheld;
An elephant from one of those
Two mighty armies is broke loose,
And with the horror of the fight
Appears amazed, and in a fright;
Look quickly, lest the sight of us
Should cause the startled beast t'imboss.
It is a large one, far more great
Than e'er was bred in Afric yet;
From which we boldly may infer,
The Moon is much the fruitfuller.[12]

'Imboss' means to take shelter; but alas, the savant had mistaken a mouse in the telescope for an elephant on the Moon: and it was indeed easy, and still is in science, to confuse appearances generated by instruments with real phenomena of nature. We owe Hooke our respect not only for his pictures, but because, despite having an instrument that was very imperfect by later standards, he was a much more careful observer than Butler's philosopher.

Hooke's work was soon followed up, in Britain, Italy and the Netherlands, as the promise of microscopic vision drew in those sharing Aristotle's concern with the generation of animals and plants. Nehemiah Grew (1641–1712), a physician from Coventry with a Leiden doctorate, was a protégé of Hooke, who proposed him for the Royal Society in 1671; he came to London, and the two men were its joint secretaries from 1677. Grew published a catalogue of its museum (a cabinet of curiosities, since dispersed) in 1681, with handsome plates; but his major work was *The Anatomy of Plants* (1682), in which the most famous illustrations are microscopic sections through branches of trees. These were not superseded until the nineteenth century, being reused, for example, in Davy's *Agricultural Chemistry* (1813) – by 1830 much better images could be obtained. Comparative anatomy was an established part of zoology going back to Aristotle's interest in homologies, and of medical science; but for plants, it was new and the microscope made it possible. Grew was a pioneer in recognising the sexual nature of plant reproduction: botany was and remained largely a descriptive and classifying science, but Grew helped to underpin it with understanding of functions and structures, often minute.

Another physician, Marcello Malpighi (1628–94) of Bologna, had been stimulated by Hooke's work to apply the microscope to animal structures. He investigated the lungs of frogs, the structure of the liver and kidneys, and the skin and its pigmentation. He also revealed the capillaries that link arteries and veins, thus confirming the circulation of the blood. Henry Oldenburg invited him, as secretary, to tell the Royal Society about his work; he did so, and was elected a Fellow in 1671. Quite independently, he and Grew had developed very similar interests; he moved into botany, and his Latin *Anatome Plantarum* was published by the Royal Society in two splendid tomes, in 1675 and 1679. The volumes contain not only handsome allegorical frontispieces with cherubs climbing a tree and wreathing three leopards, but also a stunning series of a hundred engraved plates of dissections. The first volume contained a supplementary treatise in which he had repeated Aristotle's experiment of opening hen's eggs to trace the development of the embryo. Armed with a microscope, he could see more; like Grew, he drew specimens both at their natural size and magnified, so that readers could see the power of his instrument in revealing details invisible to the naked eye. Also like Grew, Malpighi in later life spent more time in medical practice, becoming papal physician in 1691. But henceforward a microscope would be essential equipment for the aspiring botanist or zoologist.

Because compound microscopes like Hooke's were expensive and difficult to use, and distortions inevitable with two lenses, many preferred 'simple microscopes', which were in effect powerful magnifying glasses. Leeuwenhoek of Delft, an executor of Vermeer's will, was the founder of microbiology; he used tiny spherical lenses set in metal plates held up to the light close to the eye, and made by a process he kept secret. He made over five hundred of them, and it seems unlikely that he ground all those lenses – they were probably tiny spheres made by carefully drawing out molten glass. However it was done, they were extremely powerful: whereas compound microscopes only magnified about thirty times, Leeuwenhoek's magnified two hundred times or more. Unlike Grew and Malpighi, he was a tradesman with only elementary education, apprenticed to a draper (where magnifying glasses were used to count threads in examining cloth) and afterwards apprenticed to a merchant and city official. He spoke and read only Dutch. The Leiden-trained physician Regnier de Graaf (1641–73), working in Delft and interested in generation, was impressed by Leeuwenhoek's work and shortly before he died he wrote about it to Oldenburg. The latter contacted Leeuwenhoek and translated and published his replies. The observations reported by this provincial tradesman seemed almost incredible, but were duly confirmed by Hooke and demonstrated to the Royal Society in 1677. In 1680 Leeuwenhoek was elected a Fellow, and accordingly sent microscopes and a long series of letters, too informal to be called papers but abounding in discoveries. Hooke, exploring the world of little things almost beneath notice with the naked eye, had discerned wholly unexpected and rich details; Leeuwenhoek's brave new world, on the other hand, was full of extraordinary animalcules undreamed of before he entered it. Instruments and careful observation were generating descriptive science.

His descriptions were vivid. The following one is of what is now called a protozoon:

> These little animals were more than a thousand times less than the eye of a full-grown louse (for I judge the diameter of a louse's eye to be more than ten times as long as that of the said creature), and they surpassed in quickness the animalcules already spoken of. I have divers times seen them standing still, as 'twere, in one spot, and twirling themselves round with a swiftness such as you see in a whip-top a-spinning before your eye; and then again they had a circular motion, the circumference of which was no bigger than that of a small sand-grain; and anon they would go straight ahead, or their course would be crooked.[13]

In rainwater that had stood in a butt, in scrapings from his own teeth and other people's, in his own excrement, in the gut of a horsefly – wherever he looked, he found teeming microscopic life. There was no reason for him to connect the bacteria he observed with disease, and nobody else did so either for two centuries. Contemporaries were, however, aware of the importance of his observations on human spermatozoa in 1677, because of the light these cast on generation: there was widespread controversy over whether embryos were a new combination of elements from each parent (epigenesis) or were preformed, in the womb or (following Leeuwenhoek's work) in the sperm. Leeuwenhoek was very careful in his depictions of spermatozoa, but, given the fuzzy nature of microscopic images, it is not surprising that some saw them as homunculi with tails.

The microscopic world was perhaps even more astonishing than the infinite universe disclosed by the telescope and interpreted by Newton. But Copernicus' view was not, strictly speaking, proved; and Hooke's hope that the microscope might reveal the atoms or corpuscles that made up matter was similarly not fulfilled, so that atomism also remained hypothetical. Science is like that. But whatever metaphysicians thought, now that men could see further and deeper, the world and human prospects looked different. Experiment involving prisms and lenses, clockwork, chemical apparatus, barometers and air pumps, telescopes and microscopes, provided new and barely imagined ways of studying and then perhaps of changing the world. The Scientific Revolution began in experiment, equipment and technique as well as in a new philosophy. And running through all the optical experiments, the chymists' work, the pneumatics and the microscopy was a theology of nature: the new cosmos was a revelation of a wise creator and lawgiver, whose mind and purposes could now be investigated by those who could read the Book of Nature. William Blake (1757–1827) put it more enigmatically:

> The atoms of Democritus
> And Newton's particles of light
> Are sands upon the Red Sea shore,
> Where Israel's tents do shine so bright.[14]

It is to the interpretation of God's Book of Nature that we next turn.

CHAPTER 6

Through Nature to Nature's God
The Two Books

THE MYTH THAT SCIENCE and religion have been endlessly at war, sometimes cold but often hot, comes from the nineteenth century. Actually, science grew out of a religious impulse to understand God through His works. 'Atheism' in the seventeenth century generally meant practical disregard for the precepts of religion: nobody would want their daughter to go out with an atheist. But there were very few who actually denied the existence of God. The battles were about doctrines and authority, biblical interpretation, the status of scientific theory (which might be seen as well established, or as merely provisional and hypothetical), and the place of theology and science in the hierarchy of knowledge, and in universities. Otherwise, the major tension was between the God of philosophers or 'gentile prophets' like Plato, the remote First Cause or Unmoved Mover increasingly revealed by men of science as the creator of a magnificent, varied and law-governed world; and God the Father, the personal God of Abraham, Isaac and Jacob, and of Jesus, who answered prayer and was revealed through the history of the Jews and the teaching of prophets and apostles. At one limit, focus on the First Cause led to Deism, where God had, as it were, set the clockwork universe running and then sat back from it, in endless Sabbath rest. Such a God was to be worshipped by finding out how admirably the clockwork ran, and praised in cathedral or college-chapel choral evensongs, but could not be expected to disturb the smooth running of this best-of-all-possible worlds by performing a miracle, or otherwise responding to human importunity. At the other limit, focus upon human sin, God's love and Jesus' self-sacrifice led to pietism, where faith believes nor questions how, and science must not be allowed to divert anyone's time and attention from love and duty to God and neighbours.

The Scientific Revolution coincides almost exactly with what some historians call the Long Reformation, 1500–1750, during which Christianity was transformed. Whereas in its first three centuries conversion had been a matter of individual or family conviction, from the time of the emperor Constantine (274–337) the Church had been communal and collective, largely a matter of practice, defining a social world; conversion was a top-down process, following the lead of a duke or king. Life for everyone was marked by being brought to church as a baby at baptism, being married at the church porch and being carried out feet first to the churchyard. Meanwhile, sacraments and rituals (some of them pagan survivals) provided support and consolation in sickness, desolation and bereavement. In the sixteenth century religion became increasingly individual, even private, a matter once again of conviction and belief which might involve breaking with the established church and regular attendance at public worship in favour of smaller gatherings of the sometimes-persecuted elect. Religious authority was broken, private study and praying encouraged, and independent reading and thinking easier.

The empirical emphasis in the Scientific Revolution affected religious thinking more subtly. The crawling caterpillar, dying into its chrysalis-coffin and rising a glorious butterfly, had been a symbol of resurrection; the Moon, reflecting its light from the Sun, a symbol of dependence. A grittier way of thinking came in with serious entomology and astronomy. Webs of correspondences between the earthly and the heavenly were broken; and the Bible, available to Protestants in vernacular languages, was read in a literal way rather than as allegory, poetry and history. An important feature of the Abrahamic faiths, Judaism, Christianity, and Islam, was that the universe was a created thing: nature was 'it' rather than 'she', not something to be worshipped, but made by a craftsman-God in whose image mankind, male and female, was also created. The human intellect, a finite and constrained version of God's mind, could therefore understand (if only in part) what God had made; and mankind was also commanded to make use of it. Religion traditionally stood upon the three pillars of Scripture, Tradition and Reason; with the Scientific Revolution, Reason came to mean science rather than logic, and to take more weight. The three faiths had been 'people of the book': true religion came to be seen as readily compatible with science if reading the Bible was accompanied by reading of the Book of Nature. It was, after all, highly plausible that finding out about the world would tell us about its Maker and ours; and that the 'two books', though sometimes seeming at odds, must in the long run agree and cohere.

In 1517 Martin Luther posted his list of 'theses' on the church door at Wittenberg, and what we call the Reformation was the consequence – such posting was not outrageous in a university town, but getting the theses into print made them potent. Britons in the seventeenth century liked to see Protestantism as compatible with science, and Catholicism as opposed to it; but in fact the sciences flourished in Italy and France just as much as in Britain and the Netherlands. On a more mundane level, though, parson's sons (I am one) in Protestant countries were an important part of the emerging scientific community right through the Scientific Revolution, and indeed beyond. Poor but well educated, themselves often becoming, like their fathers, tutors, teachers, academics or parochial clergymen, they were likely to be interested in nature and nature's God.

This interest might take the form of natural theology, arguing for the existence, wisdom and benevolence of God from the Creation; or the gentler theology of nature, taking God's existence for granted and delighting in the wisdom revealed by close observation of nature, by mathematical harmonies and by crucial experiments. Such sentiments abounded in popular science, and indeed provided a respectable motive for what might otherwise seem self-indulgent curiosity. The term 'engagement', with overtones both of betrothal and battle, has been usefully proposed to describe the relationship between science and religion. Certainly, what we shall find is controversy within and between religious traditions, usually Christian, as religion became increasingly private, a matter of personal conviction rather than social practice, with 'belief' coming to imply assent to propositions rather than faith or trust in people or institutions: 'I believe in the Real

Presence of Christ in the Eucharist' rather than 'I believe in Britain'. This has meant that religious doctrine and scientific theory have, among Christians and their critics, assumed a much greater importance than practices. That would be a mistake. Everyone knows that practices and rituals are important in Judaism and Islam, defining communities; the same is true of Christian churches, and similarly common experience in laboratories, observatories, field studies and museums binds scientists together. We should in looking at both religion and science be sensitive to how people behave and socialise, as well as to what they claim to believe.

Searching through nature for nature's God was a very old enterprise, long antedating Christianity. Right through the Scientific Revolution it provided a rationale for observing, reasoning and experimenting, a route into science and a way of popularising it. But Luther's Reformation began an age of violent religious strife, when heretics and dissenters were persecuted, and people went to war over diverse and deeply held views of doctrine and church organisation. During the seventeenth century, amid wary war-weariness, the religious toleration that had characterised Rudolph II's Prague came formally to the prosperous Netherlands and then to Britain too. Though in both countries Roman Catholics and those who rejected the established church were second-class citizens, this was an extraordinary break from the long-established pattern that everyone in a state should share the same faith, and a major step towards making religious belief and affiliation a matter of private choice – one of the freedoms prized today. Atheism was beyond the pale: there were people deemed atheists – libertines at Charles II's court, for example – but actually to embrace atheism systematically was dangerous. By the eighteenth century the distant figure of nature's God, the First Cause of everything, had become a figure of rational worship for Deists for whom religion involved wonder, morality and social stability. With difficulty, this impersonal God could be made to seem also the fatherly God of Abraham. Subtle but not malicious, He was not a trickster deity, and men of science were attracted by this view of the world as a kind of self-winding watch whose mechanisms we could elucidate by observation, experiment, logic and mathematics. Some Deists had a sincere reverence for this creator-God, but such religion could be formal and perfunctory: Lord Shaftesbury (1621–83) famously remarked that men of sense are really but of one religion, but that men of sense never told what it was.[1]

In the earlier part of our period, it was very different. Men of science (and sense) were also men of faith, prepared to suffer inconvenience or much worse rather than abandon their allegiance. This does not mean that they were always orthodox, or coolly rational: scientific visionaries might, for example, share the alarmingly polar, complex world of the shoemaker mystic Jacob Böhme (1575–1624), for whom '[t]here is nothing in nature wherein there is not good and evil; everything moveth and liveth in this double impulse, working or operation, be it what it will'.[2] Natural magic with its practical objectives was an important component of the Scientific Revolution, but it aroused suspicion. All the churches were much concerned about magic, because it was assumed to be diabolical: the result of a Faustian pact with the Devil. Ficino insisted that his magic was white; Prospero's magic in Shakespeare's *Tempest* (1610–11) might look white to Miranda, but to Caliban he was a sorcerer. The experimental philosopher like Hooke inherited the magician's hope of improving the world by manipulating natural agencies; and it is and was hard to draw the line between natural magic and natural science, as Agrippa, Dee and Helmont found. As both Protestants and Catholics clamped down on folk religion as superstition, and customary ways of averting bad luck were banned, awareness of the Devil roaming abroad and seeking whom he might devour increased: the world was his kingdom, and his wiles and snares were everywhere. The wise woman and the cunning man were looked at askance, and their spells and remedies for those bewitched or haunted denounced: they cast out devils by Beelzebub, Lord of the Flies, as Jesus' enemies said he had done. Ghosts and spectres were interpreted as demons, and biblical stories of diabolical possession seemed all too real. Fear of witches became more intense, and the sixteenth and early seventeenth centuries saw a peak in witch trials. Religion and science embody practices, ways of behaving, as well as bodies of doctrine; they engaged with each other as styles, rituals and hierarchies, as means of coming to terms with life. We shall see that more clearly when we come to look at medicine.

Dee and Harriot were vehemently suspected in Protestant England, while Campanella, Helmont and Bruno all fell foul of the Inquisition,

Bruno fatally. So did Galileo. Churches could be expected to persecute Satan's allies, but the case of Galileo, a good Catholic and no magician, has given rise to the powerful myth that religion and science are essentially in conflict. Like all legal cases, his was unique. Galileo has now been rehabilitated, like victims of Stalin's Purges: but what does that mean? Lucretian atheistic atomism was a heresy, but Copernican theory was not: no crucial doctrines were threatened. Pique, and the familiar inability of governments and great institutions to admit mistakes in a timely or gracious way, meant that rehabilitation came three centuries after his condemnation. Rather than a straightforward tale of superstition and invincible ignorance putting down science, Galileo's story involved patronage and interprofessional jostling as laymen sought to displace clergy. But it cannot be simply a story about these things, because Galileo was forced to recant not his disobedience but his adherence to a moving Earth: it is unlikely that as he got to his feet he muttered that it moved. The pope who had been his friend and patron condemned him and his theory. There was nothing surprising about papal patronage, for the church had consistently been a patron of scholars, as of musicians, goldsmiths, sculptors, painters, stonemasons and woodworkers.

Science depends upon wonder and disciplined curiosity, but requires and promotes wealth and power. It grew in an intimate relationship with the rise of capitalism. The plot of Shakespeare's *Merchant of Venice* (c.1597) depends upon usury, lending at interest, long forbidden to Christians. Galileo came to fame in Venice, governed by a merchant oligarchy that by his time was relaxed about usury, glad of the prestige of the University of Padua (where theology was not taught) and delighted when Galileo's telescope promised real usefulness. To the Venetians' chagrin, and despite offers of a rise in salary, Galileo departed for Florence, ruled by a family of patrician bankers with a long tradition of cultivating the arts and sciences. But in Italy the papal court still remained the most glittering centre, and Galileo was attracted like a moth, especially as the pope found him congenial. But whether at the courts of the emperor, the kings of France and England or ducal establishments, those in service have to know and remember their place, as clients of capricious masters. They may be favourites, but are also vulnerable and easy to sacrifice – as happened to Thomas

More (1478–1535), Cardinal Wolsey (1475–1530) and then the reformer Thomas Cromwell (c. 1485–1540) under Henry VIII. Outraged by Galileo's failure to toe the line, the pope could sacrifice his tiresome, arrogant client.

Patronage, though we call it corruption when we see it happening today, was how society worked in Europe right through the Scientific Revolution. Galileo moved from the collective support of the Venetian state in Padua to the personal patronage of the Medici dukes in Florence, who continued to uphold him to the end. When he was condemned, the truth or falsehood of Copernicus' theory was not re-examined, and the 1616 ruling remained in place for centuries: Catholics like Gassendi had to teach and hold it as a hypothesis, and that was how it was presented to seminarians at Douai preparing to work more or less secretly in Protestant England. Because the line between hypothesis and theory is fuzzy, this did not inhibit physics seriously in Catholic German lands, in France or in northern Italy, where an edition of Galileo's *Works*, lacking only the offending *Dialogue*, was published in Bologna in 1656. Science did require the questioning disposition that Inquisitors detested, but the idea that it was compatible only with atheism or a puritan ethic is manifestly false.

The churches' patronage of scientifically minded clergy and of universities was much wider than the pope's personal support, and continued unabated. The Jesuits who had promoted Galileo's ideas in China as examples of Western achievement had to backtrack; and though in Bohemia they had been foremost in stamping out as heresy its previous intellectual liveliness, everywhere they continued to teach modern learning, including science, and to practise it in their schools and universities. At the University of Coimbra, there were two libraries: one for students, and the other for the fathers, which held the dangerous books that they were expected to refute. One of the Jesuits' stars was Athanasius Kircher (1601–80), who had studied at their college at Fulda in Hesse and joined the order. Working in France, he was in 1635 summoned to Vienna as imperial astronomer in Kepler's place. The Thirty Years' War was raging, his route to Vienna was circuitous and he was shipwrecked near Rome, where instead of going further he spent the rest of his life, at the Collegio Romano. There he could indulge his immense range of interests while training very bright students for the courts and universities of Catholic Europe, and the Jesuits' worldwide mission: he invented a speaking tube and a magic lantern, directed a museum, studied and published on music, fossils, languages and codes, attempted to decipher hieroglyphics and became seriously interested in earthquakes. In 1637–8 he visited Malta: the volcanoes Stromboli and Etna were erupting, and when his party stopped at Naples he had himself lowered into the crater of Vesuvius, which was also rumbling, risking like Pliny the fate of a martyr to science the better to see what was going on. He recorded his observations, and in 1665–6 published *Mundus Subterraneus*, with a celebrated engraving of a section through the Earth showing its fiery interior. The book aroused wide attention, with an abridged English version published in 1669. It was possible to be both 'Renaissance man' and Jesuit priest.

Nevertheless, to Protestants Galileo's condemnation provided useful ammunition in controversies, and indicated that his ideas were probably true; churchmen in Britain resolved to avoid stances that would drag them into conflicts over empirical matters and mathematical expertise. Though anticlerical feeling is nothing new, and clergy could be suspected (even more than fellow professionals in law and medicine) of having to toe a party line, they were everywhere prominent in teaching, which combined easily with a clerical career, and some were enthusiasts for the new science, nowhere more so than in Britain. One bishop, John Wilkins, was the moving spirit in the early Royal Society, and another, Thomas Sprat, its first historian/apologist. Bacon, promoting the Book of Nature as an accompaniment to the Bible, popularised the idea that God was to be understood as much through His works as through His Word, now available, at least to Protestants, in vernacular languages but by no means as straightforward to comprehend as Reformers had hoped. The Hebrew Bible showed God at work in human history, and that work had been carried forward by the church and Christian states; but He worked in natural history too. Within the Bible, this view received support: creation stories, the Psalms, the Wisdom books, prophecies and apocalyptic books like Daniel and Revelation all pointed to the value of looking for evidence of God's wisdom, goodness and activity, as well as for signs of the times. Science was thus momentous, a serious activity rather than a mildly eccentric hobby or exercise in curiosity: it could give us glimpses of Heaven.

Jesuits, friars and Augustinian canons were not, like monks, confined to the cloister, and were much involved in teaching at all levels, notably in universities; but in monasteries also the allotted study periods could include natural sciences, and teaching be part of their mission. All such lives gave opportunities for leisure, study and research, but involved celibacy; as did life as a chaplain or abbé in France. In England, while the Professors who lectured spasmodically on particular topics that often had little connection with the syllabus were allowed to marry, the Fellows of the colleges at Oxford and Cambridge, the Masters who did the bulk of the teaching, had to be celibate. The Fellows were thus either young men who had recently graduated, or confirmed bachelors enjoying port and high-table gossip, 'monks' as the historian Edward Gibbon (1737–94) derisively called them. This system lasted right down to the mid-nineteenth century – it is not perhaps so different from what in effect happens in most universities today, when tenured posts are hard to come by. Fellows were normally required to be ordained as clergymen; and most of them would hope that patronage from family, from the college or from a former pupil would bring them the opportunity of a 'good living', a well-endowed parish, where they could marry and settle down. This system of ecclesiastical patronage generated not only the familiar figure of the parson-naturalist, but also parson-scholars working in astronomy, chemistry and mathematics. But it was patchy and, as well as disappointed Fellows among the clergy, there were poor curates, never able to find security. Nonconformist ministers and laymen, including Quakers who had no clergy, received a more modern education in the Dissenting Academies which they set up than their Anglican contemporaries: such ministers were paid by their congregations and might expect less comfortable lives than the more fortunate among the Anglicans, but many shared a love of science and practised it.

Sometimes the two books that they studied did not seem to be telling the same story. The much admired St Augustine had been unable to take the first chapter of Genesis, which has days and nights happening before the Sun was created, as a straightforward account of what happened in a single week. The important part of the story for him was that God created the world, made men and women in His own image, and saw that it was good. More science brought further incompatibilities: in the Bible, the Earth is flat and the Sun circles it, things impossible to believe in the time of Magellan and Galileo. For Bacon, such contradictions were a feature of our finite minds and incomplete knowledge: in the long run they would turn out to be apparent rather than real. When, as Bellarmine had noted, literal meaning of Scripture was empirically disproved, it could be reinterpreted allegorically: after all, much of the Bible is poetry and parables, and commentators had always looked below the surface for the meaning and fulfilment of prophecies, for recurring 'types' and for arcane understanding of stories that seemed banal or unedifying (carried to an extreme of numerological decoding in Cabbala, highly suspect to orthodox Jews and Christians). Augustine's doctrine of accommodation, taken up again by Galileo, was very important in understanding how the Bible could be an infallible guide and yet apparently in error about matters of fact. God had not sought to teach us physics or biology by Revelation: He gave us energy and godlike intellects to guide us to such knowledge. When revealing religious truth, He accommodated His message to the general understanding of the time: something that our reading of the Book of Nature would correct and improve. Though it may be confused with it in polemic, this is not the same as a liberal-minded belief that we should accommodate the Bible's messages to the Zeitgeist.

Eventually, it was hoped and expected, the two books, each illuminating the other, must cohere because in concert they disclose one God, the Creator and Preserver: Baconian reading of the Book of Nature appealed to those weary of religious disputes because it seemed that it was an enterprise in which Catholics and Protestants, Episcopalians, Independents, Calvinists, Lutherans, Baptists and Quakers could all join and agree. Natural theology means arguing for the existence, wisdom and goodness of God from the world, and the empirical argument from Design became central to it, displacing logic-based arguments from causality, or ontological necessity (as in Descartes). A certain optimism is needed, a focus upon the smiling face of nature, to make the benign but distant God of nature fatherly, caring for the individual; but, certainly, contemplating the starry heavens, their motions harmonised by simple mathematical laws, and the anatomy of humans or of fleas would readily make one infer a Designer.

There were those, however, who preferred to start from the Bible. The prodigiously learned Irishman James Ussher (1581–1654) was one of the first alumni of the newly founded Trinity College, Dublin, the first Irish university. He became a distinguished Fellow, then vice-chancellor of the College, and in 1625 archbishop of Armagh. He was famous in his own time for his support for the autonomy of the Anglican Church of Ireland, and for its distinctly Calvinist theology. In his historical works he portrayed it as a direct descendant of the Celtic Irish Church, and was severe in his strictures on the Roman Catholic Church, which retained the allegiance of most of the Irish. Exiled in England during the Civil War and Commonwealth, he wrote *Annales Veteris Testamenti* (1650–4), a work of immense scholarship in which he sought to fit dates from Jewish and other histories into one chronology. Out of this came his famous date for the Creation of Adam and Eve, 23 October 4004 BC. Because of the respect in which he was held, this date came to replace others based upon slightly different biblical genealogies; and though it was never part of official doctrine, for over 150 years almost all English-speakers (and many in continental Europe) took it for granted that the world was less than six thousand years old.

Though this then seemed a very long time, it was undoubtedly a constraint. We should not, however, suppose that chronology was a bee in Ussher's bonnet: it was a highly regarded activity as serious study of history came to seem valuable and illuminating. Newton (equally learned perhaps in church history) worked throughout his life on an unorthodox magnum opus that he kept to himself, but in 1728 a fragment of it, his *Chronology of Ancient Kingdoms Amended*, appeared. There, as well as working out the ground plan of Solomon's Temple, he used astronomical data to correct but not utterly overturn earlier datings of events from ancient history sacred and profane, such as the voyage of the Argonauts in search of the Golden Fleece. The vision of history on the Hereford map was being replaced by one with far more events, ancient and modern, on a timescale. A linear Newtonian vision of time supplemented, even supplanted, the ever-circling years of the church's and the farmer's calendar, and the almanac; it would go with wage labour, clocking in and a new world where time was money. History became more significant, a process (as in the Jewish Scriptures) rather than a search for precedents or exemplars. Many of Newton's contemporaries shared Charleton's antiquarian and scientific interests, writing civil and natural histories, and when the Society of Antiquaries was founded in London in 1717 its membership overlapped with that of the older Royal Society.

If a 'young Earth' was one consequence of Bible study, another was emphasis upon the Deluge, Noah's Flood. A contemporary of Newton, Thomas Burnet (c.1635–1715) studied at Christ's College, Cambridge, where the Master, Ralph Cudworth (1617–88), was a Platonist deeply influenced by Descartes. Burnet became tutor to the grandson of the duke of Ormond, and in 1685 Master of the Charterhouse, a London school founded in what had been a Carthusian monastery. There, in 1687, he resisted James II's (1633–1701) attempts to appoint a Roman Catholic to a teaching post, and after the Revolution of 1688 was duly rewarded with positions at the court of William III (1650–1702) and Mary II (1662–94). In 1681 he had published the Latin text of his *Sacred Theory of the Earth*, which he translated into English in two parts dedicated to Charles II and to Mary respectively. His stately style in both Latin and English won him many admirers, if few converts.

The theory depended upon the doctrine of accommodation, which he had discussed with Newton. It was based on his reading of Genesis, and could be said to parallel for geology what Paracelsus and his followers had done in chymistry. Moses, supposed to be the inspired penman of Genesis and the other books that open the Bible, the Pentateuch, had merely sketched the process: Burnet sought, with Cartesian ideas and original thinking, to fill out the picture. As the chaos subsided in the Creation, the earth, being denser, sank beneath the water; but on top, like the scum in a stockpot, floated a layer of fatty rich soil forming a level surface upon which the Garden of Eden was placed. The Earth's axis was not inclined, so there was no vicissitude of seasons; vapours arising condensed at the poles, where rivers flowed towards the equator, but there was no rain (and hence no rainbows) where antediluvian mankind lived. The Earth was, as Descartes had suggested, not a perfect sphere, but a prolate spheroid flattened round the equator.

In Noah's time, miraculous heavy rain cracked the rich shell, the fountains of the abyss opened and life outside the Ark was destroyed. The

Earth's axis tilted, so the climate deteriorated and humans no longer enjoyed long lives like Methuselah's because of the vicissitude of the seasons. As a result, instead of a primeval paradise, we live upon a ruin, where disfiguring mountains are the remnants of the shell and the disturbed earth beneath the waters:

> All that we have hitherto observ'd concerning Mountains, how strange soever and otherwise unaccountable, may easily be explain'd and deduc'd from this original; we shall not wonder at their greatness and vastness, seeing they are the ruines of a broken World . . . you would easily believe that these heaps would be irregular in all manner of ways . . . and they would lie commonly in Clusters and in Ridges, for those are two of the most general postures of the parts of a ruins, when they fall inwards.

Things will be restored to something like their primeval condition by fire at the end of time. Defending himself against the claim that this was all hypothesis, Burnet continued: 'When things are either too little for our senses, or too remote and inaccessible, we have no way to know the inward Nature, and the causes of their sensible properties, but by reasoning upon an *Hypothesis*.'[3] Hooke, engaging with Burnet and rejecting his hypothesis, was much more empirical-minded, and like Kircher attributed great importance to central heat and earthquakes. The idea that the Earth is a ruin, fallen through human sin from its primeval perfection in the Garden of Eden, remained appealing to evangelicals, who saw it as a gulag or Botany Bay, a penal colony or sphere of probation for sinners, where pain and death were deserved consequences of disobedience rather than Design.

Another attempt to use Genesis as the clue to geology was made by John Woodward, an eminent physician, antiquary and Fellow of the Royal Society. He believed that at the Flood rocks were broken up, and shells and bones mixed with them, in the turbulence; and that as the waters receded, so everything settled out in layers, the heaviest at the bottom. His *Essay toward a Natural History of the Earth* (1695), with its account of the Deluge, stratified rocks, fossils and what others had written, is – like his career – an odd mixture of the ancient and the modern: he was famous also for his discovery of a shield that he believed to be Roman, and for fighting a duel

1. Cardinal Pierre d'Ailly, *Concordia astronomiae cum theologia* (Augsburg, 1490). An attempt to distinguish false from true astrology. The frontispiece shows astronomy and theology in conversation beneath the planetary spheres.

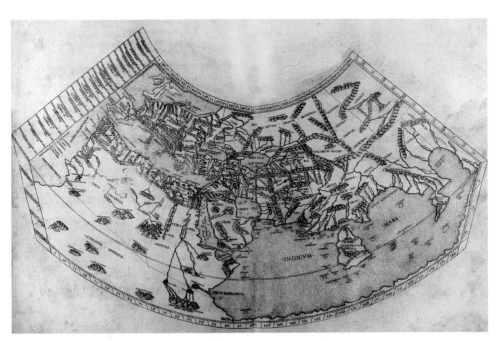

2. Ptolemy, *Geographia* (Rome, 1490). World map on a conical projection. The Mediterranean is clear, but for more distant regions he lacked good data: Sri Lanka is enormous, India small, and Scotland distorted.

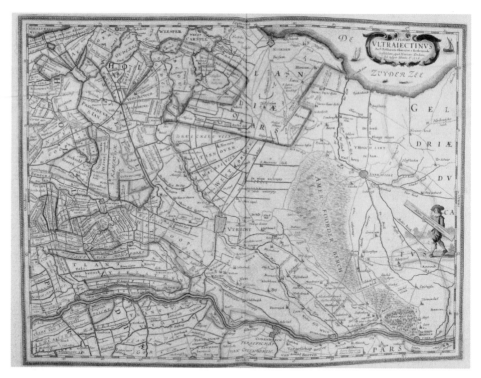

3. Gerardus Mercator, *Atlas* (1638). This map shows all the dykes in a part of Mercator's native Netherlands, around Utrecht. His world maps, with straight-line latitudes and longitudes, made navigation easier.

4. Thomas More, *Utopia* (1516). A map of Utopia, and a page of its language and script.

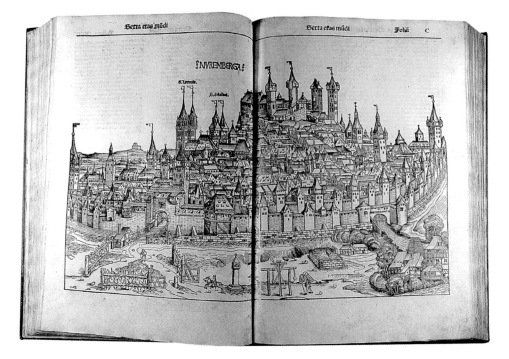

5. Hartmann Schedel, *Liber chronicorum* (Nuremburg, 1493). Nuremburg, a medieval walled city, vulnerable to cannon-fire.

6. Robert Record, *The Castle of Knowledge* (1556). Its splendid title-page says it all; this was a compendium of arts and sciences.

7 Gregor Reisch, *Margarita philosophica totius rationale naturalis & moralis principia dialogice duodecim libris cmplectens* (Freiburg, 1503). The first edition of what has been called the first encyclopaedia covering the range of medieval knowledge in all fields. The arrangement is systematic rather than alphabetical. With many woodcuts and diagrams, including this one of arithmetic showing Arabic numerals and an abacus in use.

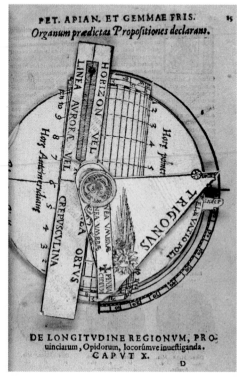

8. Petrus Apianus and Reinerus Gemma, *Cosmographia, sive Descriptio universi orbis* (Antwerp, 1584). The moveable paper volvelles to make calculations, as on an astrolabe.

9. Francis Bacon, 1st Viscount St Albans, *Instauratio magna* (London, 1620). The prospectus for the seventeenth-century revolution in scientific and philosophical method. The ships on the title-page are sailing through the Pillars of Hercules into the ocean of undiscovered truth, and returning freighted with knowledge. The arms of Christopher Columbus, and now of Colombia, have the same picture.

10. Christoph Clavius, *Gnomonices libri octo* (Rome, 1581). A book on sundials by the Jesuit mathematician chiefly responsible for the Gregorian Calendar, showing elaborate mathematical constructions involving circles.

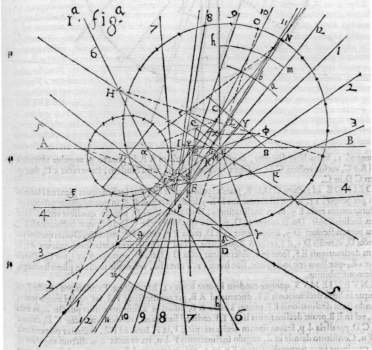

RECENS HABITAE. 23

dentalis proxima min. 2. ab hac vero elongabatur oc-

Ori.   * ○ *   * Occ.

cidentalior altera min: 10. erant præcisè in eadem recta, & magnitudinis æqualis.

Die quarta hora secunda circa Iouem quatuor sta-
bant Stellæ, orientales duæ, ac duæ occidentales in

Ori.   * ○ *   *   Occ.

eadem ad vnguem recta linea dispositæ, vt in proxima figura. Orientalior distabat à sequenti min. 3. hęc
verò à Ioue aberat min. 0. sec. 40. Iuppiter à proxima
occidentali min. 4. hæc ab occidentaliori min. 6. ma-
gnitudine erant ferè equales, proximior Ioui reliquis
paulo minor apparebat. Hora autem septima orientales Stellæ distabant tantùm min. 0. sec. 30. Iuppiter

Ori.   ** ○   * *   Occ.

ab orientali viciniori aberat min. 2. ab occidentali verò sequente min. 4. hæc verò ab occidentaliori distabat min. 3. erantque æquales omnes, & in eadem recta
secundum Eclypticam extensa.

Die quinta Cœlum fuit nubilosum.

Die sexta duæ solummodo apparuerunt Stellæ me-

Ori.   * ○ *   Occ.

11. Galileo Galilei, *Sidereus nuncius magna, longeque admirabilia spectacula pandens … quae … sunt observata in lunae facie, fixis innumeris, lacteo circulo, stellis nebulosis, apprime vero in quatuor planetis circa Iovis stellam … circumvolutis …* (Venice, 1610). Galileo's account of his discovery, using his telescope, of the surface of the moon, the nature of the Milky Way, and the satellites of Jupiter (shown in this illustration). This copy, like some others, lacks the separately-inserted illustrations of the moon.

12. Galileo Galilei, *Opere*, 2 vols (Bologna, 1656). Made up of parts with independent signatures, pagination and title-pages. This set does not include the *Discourses* condemned by the Inquisition: but its publication shows that others of his works were not censored, and that science did not collapse in Italy on his imprisonment. In the frontispiece Galileo calls the attention of the muses to the heavens.

13. Johannes Kepler, *Tabulae Rudolphinae* (1627). The frontispiece illustrating the history of astronomy through increasingly elegant columns in the temple. At the top, the Imperial Eagle distributes largesse, and at the base, Tycho's observatory and printing press on Hveen is depicted.

HYPOTHESES.

Hypoth. I. *Caufas rerum naturalium non plures admitti debere, quàm quæ & vera fint & earum Phænomenis explicandis fufficiunt.*

Natura enim fimplex eft & rerum caufis fuperfluis non luxuriat.

Hypoth. II. *Ideoque effectuum naturalium ejufdem generis eædem funt caufæ.*

Uti refpirationis in Homine & in Beftia; defcenfûs lapidum in *Europa* & in *America*; Lucis in Igne culinari & in Sole; reflexionis lucis in Terra & in Planetis.

Hypoth. III. *Corpus omne in alterius cujufcunque generis corpus transformari poffe, & qualitatum gradus omnes intermedios fucceffivè induere.*

Hypoth. IV. *Centrum Syftematis Mundani quiefcere.*

Hoc ab omnibus conceffum eft, dum aliqui Terram alii Solem in centro quiefcere contendant.

Hypoth. V. *Planetas circumjoviales, radiis ad centrum Jovis ductis, areas defcribere temporibus proportionales, eorumque tempora periodica effe in ratione fefquialtera diftantiarum ab ipfius centro.*

Conftat ex obfervationibus Aftronomicis. Orbes horum Planetarum non differunt fenfibiliter à circulis Jovi concentricis, & motus eorum in his circulis uniformes deprehenduntur. Tempora verò periodica effe in ratione fefquialtera femidiametrorum orbium confentiunt Aftronomici: & *Flamftedius*, qui omnia Micrometro & per Eclipfes Satellitum accuratius definivit, literis ad me datis, quinetiam numeris fuis mecum communicatis, fignificavit rationem illam fefquialteram tam accuratè obtinere, quàm fit poffibile fenfu deprehendere. Id quod ex Tabula fequente manifeftum eft.

Satellitum

14. Sir Isaac Newton, *Philosophiae naturalis principia mathematica* (London, 1687). First issue of the first edition of Newton's *Principia*, having one diagram (on page 112) printed upside down. This copy belonged to the chemist John Dalton, a great admirer of Newton. The plate shows Newton's list of hypotheses which at that point meant axioms or data, but later in his lifetime came to imply guesswork, which he famously sought to avoid.

15. John Flamsteed, *Historiae coelestis libri duo quorum prior exhibet catalogum stellarum fixarum Britannicum … posterior transitum syderum … [Edited by Edmund Halley]* 2 vols (London, 1712). Halley and Newton saw Flamsteed, the first Astronomer Royal, as a tiresome perfectionist who would never bring his work to fruition, and published this unauthorized edition of his observations. Flamseed, furious, had 300 of 400 copies of this edition destroyed: this survivor belonged to the Bignon library in France. His improved version was published posthumously in 1725.

L. du Guernier deli. & Sculp.

HISTORIÆ COELESTIS
LIBRI PRIMI,
PARS SECUNDA;
Stellarum Fixarum Diftantias,
GRENOVICI
In Obfervatorio Regio,
Ab Anno 1676, ad Annum 1689,
SEXTANTE Captas,
COMPLECTENS.

16. W. J. 'sGravesande, *Mathematical Elements of Natural Philosophy* (1737). Newton's work caught on rapidly in the Netherlands, and this book translated from Latin into English by the Huguenot J. T. Desaguliers became a standard textbook. The plate shows optical experiments, and a diagram showing how the eye works.

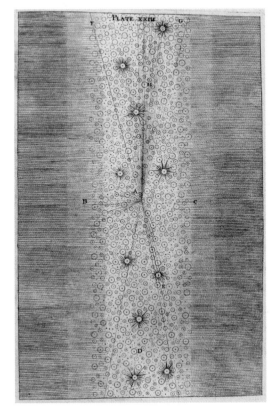

17. Thomas Wright, of Byers Green, Co. Durham, *An original theory or new hypothesis of the universe, founded upon the laws of nature, and solving by mathematical principles the general phaenomena of the visible creation, and particularly the Via Lactea* … (London, 1750). Wright's own copy, with a list of subscribers, and with annotations in his hand. In the book, Wright explains the appearance of the Milky Way ('galaxy' from the Greek) as an optical effect due to our immersion in an approximately flat layer of stars, like a grindstone: the illustration makes his point.

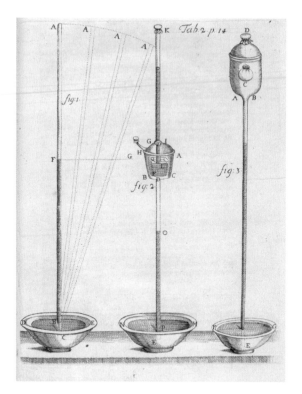

18. Academie del Cimento, *[Saggi.]*
*Essayes of natural experiments made
in the Academie del Cimento …
Englished by Richard Waller* (London,
1684). The Academie del Cimento
published experimental results as a
collective body, not assigning credit to
individuals and eschewing hypotheses.
Waller, a protégé of Robert Hooke,
was a virtuoso; a natural philosopher,
writer and editor.

19. Royal Society of London,
*Philosophical transactions:
giving some accompt of the
present undertakings, studies,
and labours of the ingenious
in many considerable parts of
the world,* vol. 1 (London,
1665). The first modern
scientific journal, publishing
peer-reviewed research
papers, almost all in English,
was at first a semi-official
publication edited by Henry
Oldenburg, secretary of the
Royal Society. This plate
illustrates an account of a
visit to China, and shows a
bridge and the Great Wall.

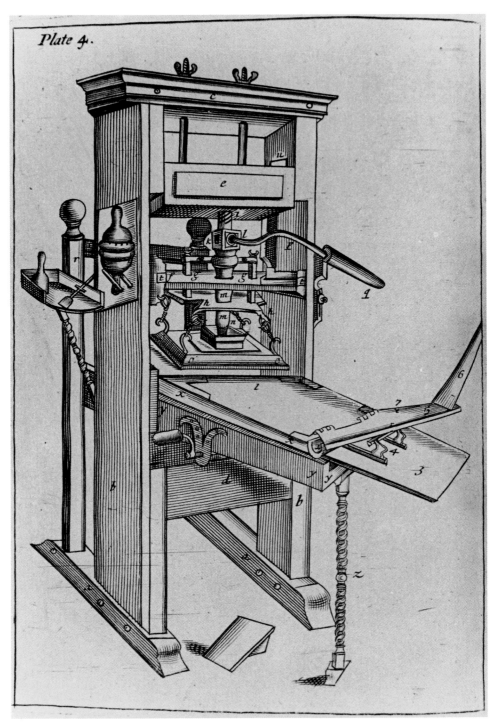

Plate 4.

20. Joseph Moxon, *Mechanical Exercises*, vol. 2 (1677–83). A printing press: crucial to the scientific revolution, making possible the wide and accurate dispersal of information.

21. *Le journal des sçavans*, vol. 1,
1665 (Amsterdam, 1679).
Le journal des sçavans (savants)
was founded by Denis de Sallo in
1665, and was loosely associated
with the Paris Academy. This
volume is from the Amsterdam
rather than the Paris edition
(which was always at risk from
censorship). Review journals,
summarising and criticising
books or recent papers, remain
an important feature of science
publishing, because nobody
can read everything: the plate
reproduces a louse clutching
a human hair, from Hooke's
Micrographia, showing how
journals spread knowledge.

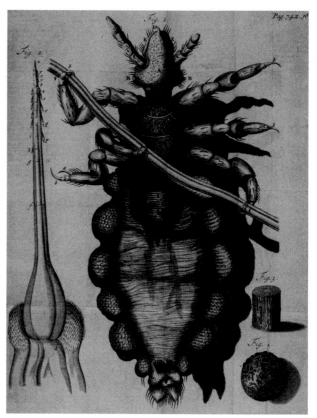

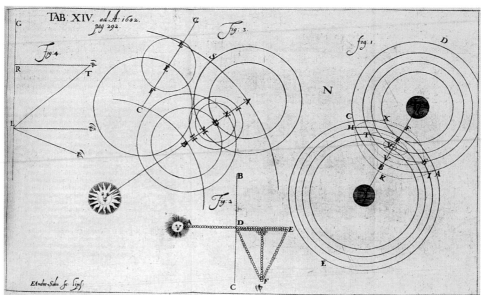

22. Acta eruditorum, vol. 1, 1682 (Leipzig, 1682). This reviewing journal embracing natural
philosophy, natural history and medicine was published in Leipzig and was known for its interest
in mathematics; illustrated are epicycle constructions.

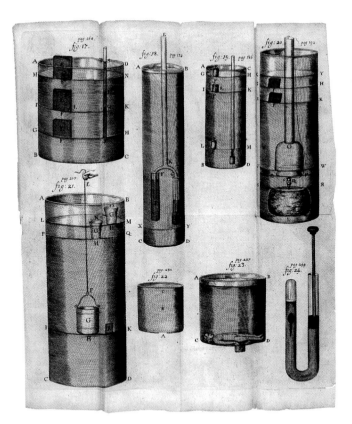

23. Robert Boyle, *Hydrostatical paradoxes, made out by new experiments, for the most part physical and easie* (Oxford, 1666). The experimental methods and observations on liquids and pressure recorded in *Hydrostatical paradoxes* helped to establish Boyle's international reputation.

A Table of the Condensation of the Air

A	A	B	C	D	E
48	12	00		29 $\frac{2}{16}$	29 $\frac{2}{16}$
46	11$\frac{1}{2}$	01 $\frac{7}{16}$		30 $\frac{9}{16}$	30 $\frac{6}{16}$
44	11	02 $\frac{13}{16}$		31 $\frac{11}{16}$	31 $\frac{1}{16}$
42	10$\frac{1}{2}$	04 $\frac{6}{16}$		33 $\frac{8}{16}$	33 $\frac{1}{7}$
40	10	06 $\frac{5}{16}$		35 $\frac{5}{16}$	35 --
38	9$\frac{1}{2}$	07 $\frac{14}{16}$		37 --	36 $\frac{15}{19}$
36	9	10 $\frac{2}{16}$		39 $\frac{5}{16}$	38 $\frac{7}{8}$
34	8$\frac{1}{2}$	12 $\frac{8}{16}$		41 $\frac{10}{16}$	41 $\frac{1}{17}$
32	8	15 $\frac{1}{16}$	Added to 29$\frac{1}{8}$ makes	44 $\frac{3}{16}$	43 $\frac{11}{16}$
30	7$\frac{1}{2}$	17 $\frac{15}{16}$		47 $\frac{1}{16}$	46 $\frac{3}{5}$
28	7	21 $\frac{3}{16}$		50 $\frac{1}{16}$	50 --
26	6$\frac{1}{2}$	25 $\frac{3}{16}$		54 $\frac{5}{16}$	53 $\frac{10}{13}$
24	6	29 $\frac{11}{16}$		58 $\frac{13}{16}$	58 $\frac{2}{8}$
23	5$\frac{3}{4}$	32 $\frac{3}{16}$		61 $\frac{5}{16}$	60 $\frac{18}{23}$
22	5$\frac{1}{2}$	34 $\frac{4}{16}$		64 $\frac{1}{16}$	63 $\frac{6}{7}$
21	5$\frac{1}{4}$	37 $\frac{12}{16}$		67 $\frac{5}{16}$	66 $\frac{4}{7}$
20	5	41 $\frac{9}{16}$		70 $\frac{11}{16}$	70 --
19	4$\frac{3}{4}$	45 --		74 $\frac{1}{2}$	73 $\frac{1}{19}$
18	4$\frac{1}{2}$	48 $\frac{12}{16}$		77 $\frac{7}{8}$	77 $\frac{1}{4}$
17	4$\frac{1}{4}$	53 $\frac{11}{16}$		82 $\frac{12}{16}$	82 $\frac{4}{17}$
16	4	58 $\frac{2}{16}$		87 $\frac{14}{16}$	87 $\frac{3}{8}$
15	3$\frac{3}{4}$	63 $\frac{15}{16}$		93 $\frac{1}{16}$	93 $\frac{1}{5}$
14	3$\frac{1}{2}$	71 $\frac{5}{16}$		100 $\frac{7}{16}$	99 $\frac{6}{7}$
13	3$\frac{1}{4}$	78 $\frac{11}{16}$		107 $\frac{13}{16}$	107 $\frac{7}{13}$
12	3	88 $\frac{7}{16}$		117 $\frac{9}{16}$	116 $\frac{4}{8}$

24. Robert Boyle, *New experiments physico-mechanical, touching the spring of the air, and its effects, made, for the most part, in a pneumatical engine* … (London, 1682). 'Boyle's law': the figures found do not exactly fit the predictions; Boyle was one of the first people to publish such uncooked material.

AA. The number of equal spaces in the shorter leg, that contained the same parcel of Air diversly extended.

B. The height of the Mercurial Cylinder in the longer leg, that compress'd the Air into those dimensions.

C. The height of a Mercurial Cylinder that counterbalanc'd the pressure of the Atmosphere.

D. The Aggregate of the two last Columns *B* and *C*, exhibiting the pressure sustained by the included Air.

E. What that pressure should be according to the *Hypothesis*, that supposes the pressures and expansions to be in reciprocal proportion.

25. Isaac Caus, *New and Rare Inventions of Water-works* (1659), showing a four horse-power gin-gan working a pump with two cylinders, converting circular into up-and-down motion.

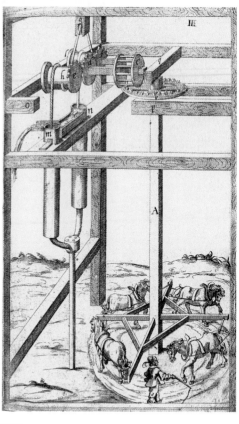

VI. Of the INWARDS.   VI. *Containing Heterogeneous Internal parts,* called INWARDS, *En-trals, Bowels, Fey, Pluck, Purtenance, Umbels, Haslet, Garbage, Giblets,* reckoning from the uppermost, may be distinguished by their Order, Shape and Uses, into

Upper; towards the Summity of the Body.

Hollow and oblong; for the *conveyance* of the || Nourishment : or of the Breath.

1. { GULLET.
{ WIND-PIPE, *Rough Artery, Weasand.*

Massie and more *solid*; within the Breast; for || *Bloud-making* : or *Breathing.*

2. { HEART, *Cordial, Core, Pericardium.*
{ LUNGS, *Lights.*

Thin and *broad*; *for partition* || *transverse*, betwixt the upper and low-er Belly : or *direct*, betwixt the Lobes of the Lungs.

3. { DIAPHRAGM, *Midriff.*
{ MEDIASTINE.

Lower; distinguishable

Both by their Shapes and Uses.

Hollow ; || *wide, but not long,* for containing and digesting of Food: *long, but not wide,* for conveying of the Food and Excrement.

4. { STOMACH, *Maw, Paunch, Ventricle, Craw, Crop, Gorge, Pouch,*
{   *Gizzard, Tripe.*
{ GUT, *Entrails, Bowels, Garbage, Chitterling, Colon.*

Massie and *solid*; for *separating* of || Choler : or of Melancholy.

5. { LIVER, *Hepatic.*
{ SPLEEN, *Milt.*

Thin and *broad* ; *by which the Guts are* || *connected* : or covered.

6. { MESENTERY.
{ CAUL, *Kell.*

By their Uses alone, as being for

Separating the Urine : or *containing* the Urine or the Gall.

7. { KIDNEY, *Reins.*
{ BLADDER, *Vesicle.*

Generation ; denoting || the *parts for Generation* : or the *Glandules for preparing the Sperm.*

8. { PRIVITIES, *Genitals, Pizzle, Tard, Fore-skin, Prepuce.*
{ TESTICLE, *Stone, geld, spay, Eunuch.*

Conception in Females, namely, the part containing the Fœtus.

9. WOMB, *Mother, Matrix, hysterical, uterine.*

26. John Wilkins, Bishop of Chester, *An essay towards a real character and a philosophical language* (London, 1668). Wilkins' plate shows his systematic classification for categorizing internal parts of animals so as to express them in his artificial language.

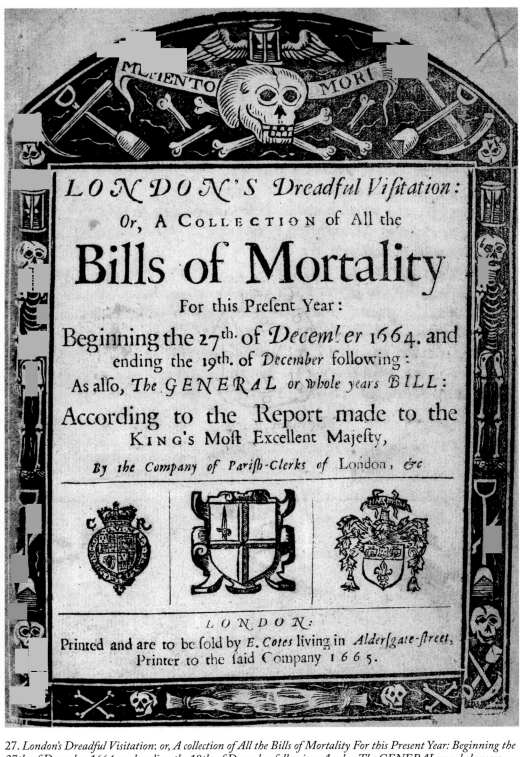

27. London's Dreadful Visitation: or, A collection of All the Bills of Mortality For this Present Year: Beginning the 27th. of December 1664. and ending the 19th. of December following: As also, The GENERAL or whole years bill: According to the Report made to the King's Most Excellent Majesty (London, 1665). The peak of mortality from the plague is shown as 7165 in the week 12–19 September. The total plague deaths for the year were 68,596, probably significantly too low a figure; deaths from other ailments show a suspiciously steep rise in the same period. The printer hoped he would never again publish so grim a report.

28. Nehemiah Grew, *Musæum Regalis Societatis* (London, 1681). Plate 1 shows specimens of alabaster, crystals and fossils from the Royal Society's very-miscellaneous collection of curiosities.

29. John Graunt, *Natural and political observations...upon the bills of mortality* (London, 1662). Dedicated to Sir Robert Moray, the President of the Royal Society, to which, on King Charles II's nomination, the author (though a tradesman) was admitted on the strength of this pioneer work on demography and medical statistics compiled from Bills of Mortality going back over fifty years.

with another doctor, of whom he said that he'd rather face his sword than his physic.

Boyle, who mulled over what is required of good and excellent hypotheses, would have agreed with Burnet about the need for them; but despite his strong religious faith he did not begin science with the Bible. Where contemporaries were coming to see a republic of letters in which experts had authority in their field, he continued to see a monarchy where theology reigned as queen of the sciences: indeed, he wrote a book about its excellence. Nevertheless, the way to settle questions in science was by observation and experiment – which to Boyle consistently pointed to mechanical explanations, and the presence of a wise and omnipotent Creator. As a good Protestant, Boyle needed no saints as patrons or inter-mediaries between him and God: he postulated an equally direct link between God and the Creation. This doctrine posed an acute problem, as it had for Böhme, over theodicy, the vindication of divine providence in the face of manifest evils. God's creation was very good: yet the young and the virtuous come to grief, and die; whole towns and villages are destroyed in storms, fires and earthquakes; poisonous plants and animals, carnivores, parasites, epidemics and endemic diseases take their terrible toll. It was a fallen world, spoilt by Adam and Eve (our representatives); nevertheless, one could not – and, in the face of what Job, the Psalmists and Jesus had said, should not – suppose complacently that the victims were guiltier than those spared.

One way out was to pick up the Platonic idea that God had entrusted the world to a subordinate Demiurge, a viceroy who did his or her best but rather botched the job; Boyle opposed this notion in his *Free Enquiry into the Vulgarly Received Notion of Nature* (1686). Nature personified was such a being, and the group of philosophers known as the Cambridge Platonists embraced her to explain why the world was less than perfect. Henry More, their founder, reacting against the puritanism of his upbringing and his Cambridge contemporaries, had corresponded with Descartes but found his ideas too mechanical. Poet as well as philosopher, in his *Democritus Platonissans* he reconciled the boundless brute-matter universe of atomists and what we call Neoplatonism, with its immortal souls and its active, life-filled, plural world:

And as the Planets in our world (of which
The sun's the heart and kernel) do receive
Their nightly light from suns that do enrich
Their sable mantle with bright gemmes, and give
A goodly splendour, and sad men relieve
With their fair twinkling rayes, so our worlds sunne
Becomes a starre elsewhere, and doth derive
Joynt light with others, cleareth all that won
In those dim duskish Orbs round other suns that run.[4]

He wrote against Hobbes, entered a controversy with Thomas Vaughan the alchemist, became a Fellow of the Royal Society in 1664, and in 1680 discussed the apocalyptic Revelation of St John with Newton. By then he had become concerned at the prevalent scepticism about witchcraft and ghosts, which for him proved the existence of spirits and the soul's immortality. Denial of the Devil's existence would soon lead to atheistic materialism:

In order therefore to the carrying out of his *dark* and *hidden designs* he manageth against our happiness and our Souls, he cannot expect to advantage himself more, than by insinuating a belief, *That there is no such thing as himself,* but that *fear* and *fancy* make *Devils* now, as they did *Gods* of old.[5]

More cooperated with Joseph Glanvill (1636–80) in investigating and popularising ghost stories, and edited Glanvill's *Saducismus Triumphatus,* first published in 1681 and going through several editions – the biblical Sadducees, opponents of Jesus, had denied the possibility of an afterlife, and the modern sceptics doubting the spirit world would soon like them deny the soul's immortality. A genial scholar, who relaxed (like Pepys) by playing the lute, More never married, refused preferment in the Church and spent his life as a Fellow of Christ's College in Cambridge. His circle turned Cambridge away from militant Calvinism to a broad-church or 'latitudinarian' Anglican tradition of tolerating, even celebrating, differences.

Glanvill also became a Fellow of the Royal Society in 1664, having written *The Vanity of Dogmatizing* (1661) promoting it. This polemic brought him into furious controversies, and he published various editions under different titles – *Scepsis Scientifica* (1664); *Plus Ultra* (1668) – as under More's influence he distanced himself from Descartes. Thomas White (1592/3–1676), a Roman Catholic priest and Aristotelian, was a temperate opponent concerned about corrosive scepticism, while Robert Crosse (1604/5–1683) and his associate Henry Stubbe (1632–76) were puritans who violently accused Glanvill of crypto-Catholic elitism, and suspected the Royal Society of the same. Glanvill's heroes (many of them indeed Catholics) proving the world was not in decline were Descartes, Gassendi, Galileo, Tycho, Harvey, More and Digby; he had no time for Aristotle, but admired the divine Plato. His sceptical and rhetorical preface set out his view of science as learned ignorance, advancing but provisional knowledge:

> The *knowledge* I teach is *ignorance*. . . . We came into the world, and we know not how; we live in't in a self-nescience, and go hence again and are as ignorant of our recess. We grow, we live, we move at first in a *Microcosm,* and can give no more Scientifical account, of the state of our three *quarters* confinement, then if we had never been extant in the greater world, but had expir'd in an *abortion;* we are inlarg'd from the prison of the womb, we live, we grow, and give being to our like; we see, we hear, and outward objects affect our other senses; we understand, we will, we imagine, and remember. . . . We breath [*sic*], we walk, we move, while we are ignorant of the manner of these vital performances.[6]

Glanvill made contact with Margaret, Duchess of Newcastle (1624–74), a philosopher, savant, poet and playwright unfairly dubbed 'Mad Madge' for her eccentricity in thus boldly entering the men's world. In 1667 she attended a meeting of the Royal Society, but it would be nearly three hundred years before the first women were eligible for election. Like More, she wrote a poem promoting atomic theory; but her coupling of mechanistic ideas with indifference to experiment made her an opponent of the Cambridge Platonists' group, which included Cudworth and John Ray

(1627–1705), the naturalist. Cudworth's magnum opus, *The True Intellectual System of the Universe* (1678), is a mighty tome, theological, metaphysical and historical, setting out a world composed of indestructible atoms of brute matter, with God its architect and nature its builder; an animated world in which life and understanding cannot arise spontaneously, atheism is impossible, and science must lead to God. Cudworth shared Glanvill's scorn of dogmatising, as in this splendid passage: 'For Things are *Sullen*, and will be as they are, whatever we Think them, or Wish them to be; and men will at last discover their Errour, when perhaps it may be too late. *Wishing* is no *Proving*.'[7] Belief in God was rational rather than projection or wish fulfilment, while to be an atheist was to be dogmatic and closed-minded.

Ussher, More and Cudworth had international reputations as theologians and philosophers, while the fame of Ray, another Cambridge man, came from his work in botany. A Fellow of Trinity College, he refused to take the new oaths required after the Restoration of Charles II, had to resign and lived in Essex near his pupil and patron the zoologist Francis Willughby (1635–72). After Willughby's early death, Ray published important works of natural history; but it was his *Wisdom of God Manifested* (1691) that made his name widely known. It launched a genre prominent especially in the Anglophone world: religious apology as a vehicle for up-to-date science. In the same year Boyle died, making a bequest for annual lecture-sermons on just that topic. Ray, forbidden to officiate as a clergyman but unwilling to become a Dissenting minister, introduced his book with a show of humility: 'For being not permitted to serve the Church with my Tongue in Preaching, I knew not but it may be my Duty to serve it with my hand in Writing.'[8] This very readable book had in fact begun as sermons preached many years earlier at Trinity College, and is thoroughly permeated with theology as well as natural history. But his vision is genially broad-church, and irenic. Faced with the variety of plants and animals, he exclaims: 'What can we infer from this? If the number of Creatures be so exceeding great, how great nay immense must needs be the Power and Wisdom of him who formed them all!'[9] The accessible little book conveys a great deal of natural history, and became an attractive, authoritative model for the subgenre called physico-theology (astro-theology dealt with the starry heavens), which also included the work of William Derham (1657–1735), parson and expert on wasps, and culminated in William Paley's *Natural Theology* (1802) and the *Bridgewater Treatises* of the 1830s.

Such writings were not confined to the English-speaking world: in France, Noël-Antoine Pluche (1688–1761), styled (as a priest without a parish) 'Abbé', published his multi-volume *Spectacle de la nature* in 1732 (translated into English in 1733). Aiming at the young, his ideal was a stimulating descriptive science: 'Transcendent Genius's' might seek to trace effects back to their causes but: 'We think it better becomes us to content ourselves with the exterior Decoration of the World and may even see that it was arraied with so much Splendor, to excite our Curiosity. Here we have Access, and . . . a Survey that abundantly fills our Senses and Imagination.'[10] The book takes the form of dialogues between a prior and two aristocrats; my copy belonged to Sir Robert Eden (1741–84), the last British governor of Maryland, who acquired it as a boy of thirteen. What he read there about the war of nature, for which insects were particularly well equipped, may have prepared him for later unpleasant experiences in revolutionary America.

While Newton is remembered for working in physics and chemistry, he was driven throughout his life by his religious faith. He wrote copiously about it, but kept his thoughts largely to himself because they were un-orthodox. As a result, these papers were little known until the later twentieth century. Newton was brought up a puritan Protestant, and retained much of this inheritance despite gradually coming to deny the divinity of Jesus and Trinitarian orthodoxy. He was strongly opposed to popery (even standing up as MP for Cambridge to the formidably browbeating Judge Jeffreys (1648–89)); was deeply interested in prophecy, the Books of Daniel and Revelation, and the identifying of the apocalyptic Last Days; and perceived comets not only as predictable heavenly bodies but also as highly significant signs of the times. In reading the Bible, he distinguished between vulgar, literal readings of the surface of the text, and true understanding arrived at through wisdom and thought. In all these things, he was not untypical of his epoch, but there was a disjunction here from the Newton admired by later generations, the exemplar of Reason, reading God's mind with his astonishing and powerful mechanics uniting Heaven and Earth.

The first set (1692) of the lectures endowed in Boyle's will was delivered by Richard Bentley (1662–1742), a classical scholar and up-and-coming cleric who went on to be a domineering Master of Trinity College, Cambridge. Bentley realised that Newton's *Principia* had provided wonderful material for demonstrating design, but the mathematical reasoning was beyond him. Boldly, he began a correspondence with Newton, published in 1756, after their deaths; the great pundit and lexicographer Dr Samuel Johnson (1709–84) declared that the papers showed the mind of Newton gaining ground upon the dark. Bentley wanted to know how far Newtonian mechanics demonstrated a wise Creator. Newton began his reply:

> When I wrote my Treatise about our System, I had an Eye upon such Principles as might work with considering men, for the Belief of a Deity, and nothing can rejoice me more than to find it useful for that Purpose. But if I have done the Public any Service this way, it is due to nothing but Industry and patient Thought.

Newton saw the distribution of matter in the universe (where one might have expected everything to congregate into one great lump under gravity), the orderly Solar System, perhaps the Earth's axis inclined to bring us seasons, and the harmony of the whole, as evidence of a Designer. In his next letter he declared that he could not believe that gravity was essential and inherent to matter, elaborating this subsequently:

> It is inconceivable, that inanimate brute Matter should, without the Mediation of something else, which is not material, operate upon, and affect other Matter without mutual Contact, as it must be, if Gravitation in the sense of *Epicurus* be essential and inherent in it. . . . Gravity must be caused by an Agent acting constantly according to certain Laws; but whether this Agent be material or immaterial, I have left to the Consideration of my Readers.[11]

God was at work constantly in Newton's universe, nature a perpetual worker, exposing Newtonians to the risk (as the orthodox saw it) of pantheism, identifying God with the world in which we carry on our lives. Leibniz objected to that, and also accused Newton of supposing God a bungling craftsman who could not make a perfect clockwork mechanism that could run without His attention. Newton and Bentley, however, were proud that their system required God's supervision, and would not accept a God who never interfered in the world, like an absent prince. Leibniz also claimed that Newton was confused about space and time. In his discussion Newton had distinguished between vulgar and sophisticated understandings; true and mathematical time, space and motion were absolute. The Earth was really, not just relatively, in motion, and space identified with God's sensorium; in God we really do live and move and have our being.[12] To the third edition of *Principia* (1726), Newton added a 'General Scholium' affirming that discourse about God certainly belongs to natural philosophy.

Newton took pains to avoid being ordained at Cambridge, because he was an Arian or Socinian, denying the divinity of Jesus; prudently he kept these views to himself. His deputy and then successor was Lucasian professor William Whiston (1667–1752), shared these beliefs, but delighted in controversy and unwisely sought to propagate and then publish them, despite being warned about the consequences. He was accordingly expelled from the university in 1710, and had to support himself thereafter in London, with help from various patrons. He had mastered at Cambridge Newton's *Principia*, and in 1696 published his millenarian *New Theory of the Earth*, incorporating ideas from Newton and Woodward; it went through six editions and was translated. He published his Cambridge lectures, and Newton's on algebra; gave the Boyle Lectures in 1707; and then turned his attention to the prophets, bringing out his *Accomplishment of Scripture Prophecies* in 1708, and the heretical *Sermons and Essays* in 1709. In London, with the instrument-maker Francis Hauksbee the Younger (1688–1763), he delivered public lectures on natural philosophy; he was also concerned about the problem of finding longitude, and active in pressing Parliament in 1714 to set up a Board of Longitude and offer a prize for the discovery of a practicable method. Very literal in his exposition of the Bible, he made a model of Solomon's Temple in Jerusalem, and lectured about it.

Bentley had reckoned that his Boyle Lectures had so thoroughly demonstrated the absurdity of their position that atheists had become Deists instead. Indeed, the God of nature became more and more credible as the architecture of the cosmos and the intricacy of animals and plants were revealed. The problem for the orthodox was that this God might replace theirs. In 1696 the lapsed Catholic Irishman John Toland (1670–1722) published *Christianity Not Mysterious*, in which he sought to show that there was nothing in the Gospel that was contrary to reason, and no truly Christian doctrine that could be properly called a mystery – a word that had ceased to designate a skill, something learned by doing it with a master, and had taken its modern meaning, an enigma. Scripture had, he believed, been tortured to countenance scholastic jargon and metaphysical chimeras, and superstition had been the result:

> Truth is always and everywhere the same; And an unintelligible or absurd Proposition is to be never the more respected for being antient or strange, for being originally written in Latin, Greek or Hebrew. Besides, a Divinity only intelligible to such as live by it, is, in humane Language, a Trade.[13]

The book was furiously denounced as Socinian and pantheistic, but Toland was taken up by Whig supporters of the Protestant succession to Queen Anne (1665–1714), and sent on a deputation to Hanover. He impressed Locke, who had written *The Reasonableness of Christianity* (1695), and Leibniz, and became one of the founders of the Whig tradition of toleration and reasonable religion in Britain.

His contemporary Matthew Tindal (1657–1733), an Oxonian and lawyer, had turned Roman Catholic during James II's reign but returned to the Church of England in 1688. His most famous work was *Christianity as Old as the Creation* (1730), which viewed the Gospel as a restatement of the religion of nature; for him, the Bible in essence told in the readily palatable form of stories, parables and poems what the wise could (and indeed had) deduced from observation. The book, like his earlier writings, was energetically criticised and parodied, and his libertine lifestyle made it easy to characterise him as both an infidel and a turncoat. In contrast, Toland and

Tindal's friend and supporter Anthony Collins (1676–1729), who published his *Discourse of Free-Thinking* in 1713, was respected as an upright country gentleman and justice of the peace, a learned man whose excellent library was rich in works by continental authors. Though denounced by the strictly orthodox, his writings were widely taken up in Britain and abroad as Deism spread and became respectable in the period that we call the Enlightenment.

Sermons were indeed preached against Deism, and Benjamin Franklin heard one in London; but it backfired, just as party political broadcasts can do, and Franklin became a Deist. In London in 1725, improving his printing skills, he served as a compositor on the third edition of *The Religion of Nature Delineated* by William Wollaston (1659–1724), whose descendants over two centuries formed a dynasty of Fellows of the Royal Society. It sold well; there were further editions, and a French translation was published in 1726. The book is primarily a naturalistic system of ethics, based on a calculus of pains and pleasures (often associated with the later philosopher Jeremy Bentham (1748–1832)), and on reason as the way to discover truth, and our consequent obligation to follow reason as best we can. This kind of ethics looks not to the development of virtues but to outcomes, so in an unpredictable world (in words famously echoed by the more orthodox Bishop Joseph Butler, 1692–1752): 'That we ought to follow *probability*, when certainty leaves us, is plain: because it becomes the *only* light and guide we have.'[14] For Butler, Christian faith (and for Wollaston, Deism) was soundly based on probability – and probability, in a new world of life insurance, annuities and stock exchanges, was an important matter, just beginning to be quantified as statistics and games of chance were studied, notably by Halley and Abraham de Moivre (1667–1754).

Those doing statistics shared Wollaston's vision of an ordered Newtonian world of particles, laws of nature, and design: 'We are so far acquainted with the *laws* of gravitation and motion, that we are able to calculate their effects, and serve ourselves of them, supplying upon many occasions the defect of power in ourselves by mechanical powers, which never fail to answer.'[15] Animals and plants are subject to laws of nature just like the inanimate world, but since matter cannot by definition think, humans have immortal souls. Separated from their bodies, they will find

The mansions, and conditions of the virtuous and reasoning part [of humankind] must be proportionately better than those of the foolish and vitious. . . . Hence it follows, that the practice of reason (in its just extent) is the great preparation for death, and the means of advancing our happiness through all our subsequent duration.[16]

'Injoyments and sufferings here' will also be taken into account, but the wicked will 'be really unhappy in that state to come'.

Arians and Socinians began in the mid-eighteenth century to coalesce into a new denomination, Unitarians, with its own ministers, notably Joseph Priestley (1733–1804); they considered themselves to be the true Christians, liberated from myth. Urban, attractive to modernisers and much involved in science and education, they were odious and frightening to the hierarchies of orthodox Churches, appalled by Priestley's incendiary compound of politics and faith. Deists, on the other hand, never formed a Church, remaining a loose group of enlightened thinkers not doubting that the clock-world had a clockmaker, liking church music, sharing a kind of practical stoicism and believing that churches were an essential social cement – particularly for restraining the lower orders.

Clearly, those who shared a rational and unmysterious faith in a distant but benevolent deity had no problem in pursuing science. Indeed, they might well feel that they had good reason and even a duty to do so because God had made it comprehensible and improvable. But not everyone found it so easy, and it was possible to see science as a distraction, a temptation to escape from the human predicament. Just as court wits saw natural philosophers as childlike and laughed mightily at the Royal Society attempting to weigh air, so the serious-minded, both Catholic and Protestant, might come to see it as a hobby, a childish thing to be put aside in a truly grown-up life. Blaise Pascal and the Dutchman Jan Swammerdam (1637–80) were two such men. Pascal was a lawyer's son; his mother died when he was four years old, and he was brought up and educated at home with his sisters. The family moved from Clermont to Paris in 1631, until his father was appointed intendant of Normandy in 1639 and they moved to Rouen. Young Blaise's aptitude for mathematics was astonishing, and before he was sixteen he had proved an important theorem in projective geometry.

At the age of nineteen he invented a calculating machine for doing additions, of which some seventy were made. Turning to hydrostatics, in 1647 he latched on to Torricelli's work on air pressure. In Rouen the family were converted to Jansenism, a stringent and morally rigorous Catholicism embracing the belief that without God's special grace good deeds are impossible, and a determinism making for theological pessimism – not altogether unlike puritanism, and strongly opposed by the Jesuits. In 1649 its doctrines were condemned by the Sorbonne, and in 1653 by the pope. Meanwhile in 1650 the Pascals returned to Paris, where Blaise's father died and his unmarried sister entered the convent of Port-Royal, the Jansenist headquarters. In failing health, on 23 November 1654 Pascal experienced a second conversion, to a personal faith in Christ as Saviour and in the God of Abraham, the knight of faith who believed God's apparently impossible promises, and obeyed His terrible commands, rather than of the philosophers and men of science. In the following year he retired to a secluded life at Port-Royal. There he published, anonymously, his brilliant *Lettres provinciales* (1656–7), defending Jansenism in fine style against the Jesuits, whose worldliness, casuistry and amorality shocked him; and composed the famous *Pensées* (1670), published posthumously from his notes.

Refuting the claim that 'there was never yet philosopher/that could endure the toothache patiently',[17] Pascal at Port-Royal (supposedly to divert his thoughts from toothache) had turned to the geometry of cycloids, finding the area, centre of gravity and the volume of solids generated from them. Some of his theorems he announced without proofs, as a challenge to mathematicians – taken up by Wallis, Wren and Huygens. In 1654 he had corresponded with Pierre Fermat (1601–65) about probability, and they can be seen as the pioneers in quantifying chance and bringing it within the domain of law, so important in the eighteenth century. To the *philosophes* of the latter era, it seemed absurd that someone of such ability in mathematics and physics should have turned away from it towards religious practice and theology, but this shift would have seemed much less amazing, indeed natural, to Pascal's contemporaries. Indeed later generations, sympathising with such a move from understanding to reason, knowledge to wisdom, have much admired the pithy wisdom of the *Pensées* for which Pascal is now generally remembered.

Swammerdam was the son of an apothecary in Amsterdam who had assembled a famous cabinet of curiosities. His father hoped Swammerdam would become a Protestant minister, but was persuaded to allow him to study medicine instead. At Leiden he became a friend of Nicholas Steno (Niels Streensen, 1638–86), later famous for identifying fossil sharks' teeth. After graduating, emerging as a very skilled dissector, he went to France and befriended Melchisédech Thévenot (1620–92), traveller, savant, inventor of the spirit level (1661) and author of a widely read book on swimming. In France, Swammerdam began his series of microscopic dissections of insects, following their life-cycles and metamorphoses. Back in Amsterdam, he had a brass-topped table made, with moveable arms to hold things in place for his microscopic work, and continued his work on comparative anatomy. His impatient father wanted him to get a proper job, they quarrelled, his allowance was cut, and he became unhappy and unwell. From 1673 he became a follower of Antoinette Bourignon (1616–80), an enthusiast who, having tried to found a religious order, rejected organised religion and saw herself as the apocalyptic 'woman clothed with the sun' (Revelation 12: 1–2). She persuaded him that science was vain, so most of his work remained unpublished. He left his papers, which included magnificent drawings, to Thévenot, whose attempt to get them translated from the Dutch and printed came to nothing. Much later, they were discovered and bought by the famous Boerhaave of Leiden medical school, which had them published in Latin as *Biblia Naturae* (1737–8), usually rendered as the 'Bible of Nature' but the English translation was titled *The Book of Nature* (1758). This splendidly illustrated folio is a classic in the history of entomology and insect development.

These two stories remind us that in the seventeenth century questions about Augustinian theology, predestination and determinism, the grace of God and the efficacy of good works, were crucial, and highly appropriate for serious study and thought; and that the end of the world seemed to many just as imminent as it did more recently in the darkest days of the Cold War. Sober thinking about one's end was in order, indeed urgent; 'Vanity' paintings were in vogue, and skulls or corpses in winding sheets featured on gravestones and monuments. Science in that perspective could easily seem frivolous. This was a world-view dominated by Adam and Eve's Fall, by sin, death and alienation, where salvation depended upon acceptance of Jesus as Saviour. But for many, cheerfulness kept breaking in, and most of those engaged in the sciences were more optimistic, filled with wonder at the world and all its variety and apparent design, rejoicing in finding laws of nature, and in the promise that knowledge was power and that by reducing toil and pain it would bring happiness. Those who thought like this took for granted that doing science made sense because God who made the world had endowed us with minds to understand it: it was difficult but not impossible, and doing it was a form of worship – though the gulf Pascal perceived and crossed between the God of the philosophers and the Father God of Abraham remained, and remains, hard to bridge.

Firm religious convictions were an essential part of early-modern science, because belief in a wise Creator guaranteed an ordered world that could be understood thanks to humans' God-given intellect. And since the earliest calendars had been drawn up, that order had been demonstrated. The alternative seemed to be a world of random occurrences. Faced with the choice between design and chance, as the Psalm has it, 'the fool hath said in his heart: There is no God',[18] or, as Edward Young (1683–1765) more elegantly put it, coming to terms with the gloom of bereavement in his celebrated *Night Thoughts* (1742–4):

> Devotion! Daughter of Astronomy!
> An undevout astronomer is mad.[19]

Science was not secular or value-free; it meant reading God's Book of Nature. The pious, rather solemn Young had met Voltaire and been shocked; but even Voltaire was a Deist. One might have thought that by the mid-eighteenth century there was enough momentum within science for it to declare independence from its religious background; but outside small circles of the advanced or dissolute, theology of nature retained its sway and atheism its stigma. The fight for independence and an equal place in the Sun came later, in revolutionary France and Victorian Britain. Even sceptics saw religion as a vital social glue. Experience as a churchwarden going back forty years has taught me that creeds and doctrines remain much less important to lay churchpeople than outsiders suppose, beliefs being variable and inconstant, and practice, with faith (meaning trust and loyalty), holding the community together. But clearly, in our period, the Reformation made an enormous difference to the way in which nature was perceived, indirectly boosting the Scientific Revolution among both Protestants and Catholics.

Nature had previously not been a book to be read but a sacred sphere with holy places sanctified by connection with a saint or with providential visions, signs and wonders. High places, often consecrated long before Christianity came to Europe, holy wells dressed with flowers in annual festivals, and shrines where saints' relics were preserved were revered. Important centres like Rome, Compostela, Canterbury and Durham were richly endowed by the gifts of pilgrims, and foci of what was in effect a tourist trade. In the country, on Plough Sunday in February, farm implements would be blessed before sowing. Then, each spring on the Sunday before Ascension Day, Rogation processions would beat the bounds of parishes, blessing the fields and praying for good crops – as well as boosting parochial self-confidence in the face of neighbours. Intercessions would be addressed to God through the patron saint to whom the local church was dedicated, and if God's favour brought a good harvest, that would later be celebrated and first fruits given to the church. If not, it would be time for heart-searching and repentance as well as belt-tightening. In times of sickness, God had given us clues to which herbal remedies would (if it was His will) heal diseases, and they should be sought. Diabolical influences and witchcraft could usually be dealt with by exorcisms, sprinkling with holy water and true repentance: the Devil had after all been defeated at Jesus' Crucifixion and Resurrection, and thanks to the Church and its rituals his power was broken. The world was permeated by God's spirit; sculptors decorating churches, and miniaturists illuminating the margins of religious manuscripts, delighted in nature and its fecundity.

To the Reformers, this seemed superstition and ignorance: pagan survivals. In their zeal, they sought to stamp out manifestations of folk religion, and in this campaign they were in due course joined by Roman Catholics, after the Council of Trent (1545–63). Protestants, especially the more fervent, went further: relics, images, vestments, stained glass and illuminated manuscripts were idolatrous and must be destroyed; and to this iconoclasm they added destruction of holy places, wells and shrines, for God was present everywhere and no part of nature was more sacred than any other. Processions, saint's-day holidays (even Christmas) and the use of holy water were anathema: we should be sober and vigilant, because our adversary the Devil, far from being defeated, walked about as a roaring lion seeking whom he might devour.[20] Evil spirits, perhaps posing as ghosts, would tempt everyone. Witches and necromancers who had yielded to these temptations were to be sought out and punished as the Devil's agents – if they were not, or if blasphemers were tolerated, God would punish the nation for its laxity.

In Henry VIII's England, after 1535 monasteries and nunneries were abolished. A few of their buildings were turned into cathedrals or parish churches, but most were sold off with the condition that they must be unroofed so that they could not be used by crypto-Catholics. Treated as convenient sources of cut stone for building, they rapidly turned into ruins. A largely married clergy – some of them former monks who might, like Luther, marry a former nun – replaced priests who had all been celibate, notionally at least. By the end of the sixteenth century, when monks, friars, canons and nuns had passed beyond living memory, antiquarians began to take an interest in monuments and texts from the past that illuminated history, often local. One could feel nostalgia for a world that was lost in those bare ruined choirs where late the sweet birds sang.[21] By the eighteenth century, they were the epitome of the picturesque, and taking melancholy pleasure in ruins, perhaps even building pseudo-ruins, became fashionable. We are used to seeing carefully conserved ancient monuments, but our ancestors had to scramble over ruins overgrown with moss, ivy, brambles and saplings.

They were also concerned with older monuments, long and round barrows, Roman remains and standing stones that aroused enormous speculative interest in the mysterious Druids. In 1956–7, I was stationed at the artillery base at Larkhill, which adjoins Stonehenge, where at the solstices and equinoxes 'Druids' would alight from a bus and change into flowing white garments to enact ceremonies under the irreverent eyes of the military. They still do, thus maintaining in practice and ritual the speculations of Stukeley. Not only standing stones, but also the barrows there and elsewhere intrigued antiquaries, who would collect friends to assist in their excavation.

Sympathetic if sentimental enthusiasts cleared out and investigated holy wells too, some of which would later be resurrected as spas.

Antiquarian curiosity, local pride, and getting out and about went very readily with the pursuit of natural history, done in the spirit of John Ray, and meant seeing nature not as inherently sacred in whole or in parts, but as God's creation in which His wisdom and goodness were manifested. Christianity and science were deeply interfused, making science widely attractive and elevating it above mere curiosity and vulgar utility. And as society became increasingly urban, so the countryside came to be seen in a new light: no longer a workplace, it was where one looked for roots, for a less artificial lifestyle, even for the Gothic architecture despised by devotees of the Baroque and Rococo. And, like the raw materials of chemistry, it could be improved. With the fashionable English style of landscaped gardens including ornamental but natural-looking lakes, exotic plants and trees carefully clumped, interest in natural history increased, supported by handsomely illustrated publications.

Neither astronomy nor natural history could thus be thought of as wholly secular; nor were they value-free. Science began to acquire a code of ethics independent of religion, though initially based upon it. Academies, and societies associated with courts, assumed a gentlemanly code of honour, in which truth-telling and careful judgement were prominent, and surprising phenomena could be demonstrated and discussed publicly. The new science had its values: dismissing prejudice, rejecting mere authority in favour of experiment, calculation and inference, looking afresh with open eyes at things that had been taken for granted, being prepared to abandon a promising hypothesis that failed when tested, taking pains to convince one's peers by demonstration and argument, being humbly ready to learn from craftsmen, gardeners and other social inferiors. Naturally, some practitioners fell short of these ideals, but the idea of science as objective, as tested public knowledge, was in place by the early eighteenth century. Everyone doing science was nevertheless guided by notions of what was and was not possible, and by the maxim that one should not trouble about investigating the impossible. Witchcraft, gold-making alchemy and astrology, taken seriously by seventeenth-century men of science, came to seem so implausible to adherents of the new philosophy that they simply rejected them, whatever evidence might be produced. We should, however, be careful about mocking our older ancestors' devotion to astrology, given the solemn faith we put in the constantly falsified predictions of economists.

Making war against abstractions like terror is problematic, and for two abstractions like religion and science to go to war would be weird: we need to keep our metaphors under control. There was friction, but clearly during the Scientific Revolution men of science and churchmen (often, indeed, the same people) were not engaged in endless conflict: the pioneers of science were men of faith, not always orthodox and sometimes anticlerical. They hoped to build Heaven, the Kingdom of God, on Earth. The Reformation had brought fresh thinking and challenges to authority; the sons of Protestant clergy entered science; Jesuits reformed and modernised the curriculum of the schools and universities they controlled; and both Catholics and Protestants assailed folk religion as superstition. In particular, nature was not to be seen as sacred, but created: 'it' rather than 'she', open to study, analysis and dissection. Wars of religion, and persecution of heretics and dissenters, were hugely important in the rise of science: refugees (of deep faith, tested in adversity) brought skills with them, acquired new perspectives, built up critical masses in tolerant cities and disseminated new learning all over Europe. But the sciences were prominent, along with the study of history and languages, in dethroning theology, making it by the late eighteenth century one specialism among others in a republic of autonomous disciplines all competing for attention. Churches' doctrines were opened to secular philosophical investigation, their scriptures to critical interpretation like other old writings, their councils to political historians, and their creation narratives and miracles to scientific revisionism. The successes of Galileo, Descartes and Newton brought confidence in the New Philosophy, with its empiricism, linear time and causal chains. Scientific theory was provisional and progressive, and this gave a new fillip and direction to scholarship in other fields such as philology and history. Abstractions cannot be at war, but religion and science are not abstractions and have it in common that they are not just individual or private practices: they involve groups and communities also. So we turn next to the academies and societies that gave science a collective identity and an appropriate language, as the churches did with religion.

CHAPTER 7

SHARING THE VISION
SCIENTIFIC SOCIETIES

SCIENCE IS ABOUT DISCOVERY, taking the lid off and seeing what is inside. The word 'discover' has distinct but complementary meanings, to find and to reveal, and both are required for science. There have always been lonely introverts solving problems essentially for their own satisfaction, and groups that have kept knowledge esoteric; but their discoveries do not really count as science, which must be ordered and public. It is part of any scientific method that discoverers have to convince first themselves and then others that they have found something new, curious, interesting and relevant to current concerns, and potentially fruitful: stray brute facts are, as we all find out, of disappointingly little concern to natural philosophers. Thus science is necessarily a social activity. Common sense needs to be refined, trained and organised. Monasteries and cathedral schools, and then medieval universities, brought scholars together, reviving and cultivating the liberal arts and training their members for the three professions of church, law and medicine. But by the sixteenth century the universities' formal disputations and conservative syllabuses were attracting criticism from Bacon and others; and by his time they had been joined as intellectual centres by courts, and more informally by printing houses. Coffee houses and the salons of aristocratic ladies would later supplement them further. But from the beginning of the seventeenth century, there were also – very importantly – the first societies devoted to science.

In a world in which patronage played an essential part, and where there was keen awareness of social status and position, the Scientific Revolution depended upon contact between savants, on the one hand, and practitioners, craftsmen, artisans, travellers, merchants and seamen, on the other. In universities, scholarship boys like Hooke and Newton could indeed make contact with young grandees who might (as happened with Willughby and Ray, Boyle and Hooke) become their patrons; the Church had always been a ladder for the socially mobile. But just as a gulf separated the university-trained physician from the surgeon who learned his trade by apprenticeship, so in other arts and sciences the theorist and the practical man were separated by a social barrier. They even spoke different languages, for Latin was the language of learning, making contact across political frontiers easier for the educated than communication with their local peasantry. This gulf that separates the cosmopolitan learned from lay people is still visible at international conferences and student gatherings, though their learned language is now English. The disdain felt in Antiquity for manual work, shared by Renaissance aristocrats, was a block to science. But noblemen and merchant princes were happy to act as patrons for engineers, builders, goldsmiths, painters, musicians, actors, gardeners, poets, alchemists, astrologers and philosophers; they were competitive and gained prestige from soft power in the form of lively courts that were centres of artistic and intellectual life. Outside the trammels of syllabuses, and with opportunities for informal contacts, courts could be wonderful places to be; but they could also be terrifying milieux, where the whims of princes, the scheming of Machiavellians, and external threats of war and violence generated an atmosphere of whispering, intrigue and backstabbing. Galileo and Harriot found how necessary but dangerous patronage was.

The very different world of the sixteenth-century printing house can still be entered at the Plantin-Moretus Museum in Antwerp, the handsome building with a formal garden in its courtyard in which Christophe Plantin (1514–89) founded his printing, publishing and bookselling business in 1555. On the ground floor there are still printing presses dating back to about 1600, large quantities of type in their cases, a room for the proofreaders, an office and the old shop. Upstairs, there is the type foundry, with a furnace, moulds and punches, and more type. Illuminated manuscripts collected, and books, copperplates, catalogues and ephemera printed by the firm are on display now, as no doubt they were in the past. But the building was also the house in which Plantin and then his Moretus descendants lived until they gave it to the city in the nineteenth century. Here Plantin entertained his authors; artists including Peter Paul Rubens (1577–1640), who illustrated his publications (notably, for our purposes, on botany and geography) and also painted portraits of the family and their associates; eminent tourists like William of Orange (1554–1618) and Archduke Albert (1559–1621); prominent citizens; and humanists from all over Europe. With other merchants and scholars of the time, Plantin belonged to a secret religious sect, the Family of Love, and was sympathetic to the Dutch Protestants though outwardly conforming to the Roman Catholicism of the Habsburg dominions. As the business prospered, the house grew and became a handsome patrician residence with a magnificent library in the midst of all the noisy activity; an informal centre buzzing with intellectual conversation. Antwerp and then Amsterdam, wealthy centres of intercontinental commerce that had neither court nor university, played a crucial role in the Scientific Revolution through the book trade.

Meanwhile, in fifteenth-century Florence, the Medicis had transformed themselves from prominent bankers into grand dukes and founded the Platonic Florentine Academy, a model of enlightened patronage followed by later academies directed particularly towards language, philosophy and history. In 1582 the Accademia della Crusca was founded in Florence to purify the language: 'Crusca' means bran, winnowed away in cleaning wheat. It formalised the Tuscan dialect which became the official Italian language, and was copied in France in 1635 when Cardinal Richelieu (1585–1642) founded the Académie Française for forty salaried 'immortals' to do the same for French. Academies like this were examples both of state support for learning and of state control: the informal groups of scholars congealed into academies under powerful patronage lost their autonomy and their fuzzy boundaries. With a less formidable patron, they might enjoy the best of both worlds. Thus at the age of eighteen the Roman aristocrat Federico Cesi (1585–1630), nephew of a cardinal, formed, with three friends, the Accademia dei Lincei, who lived together in his house and devoted themselves to the sciences. Disapproved of by Cesi's father, and suspected of black magic, the group dispersed; but in Naples Cesi met Giambattista della Porta and recruited him. Della Porta had taken the lynx that sees in the dark as the symbol of the man of science, and on his travels had sought out 'Libraries, Learned men, and Artificers' and 'by most earnest Study and constant Experience' had tested what he read or heard.[1] In 1611, while in Rome, Galileo was brought into the group, whose membership, in Italy and abroad, grew steadily through the 1620s. Cesi's interests were primarily in fossils and in natural history, particularly in ambiguous items like fossils and monsters; and he got together collections of specimens and drawings. The Lincei published Galileo's *Letters on Sunspots* and *Assayer*, and were to have published the *Dialogue*, but in 1630 Cesi died suddenly, and when Galileo was subsequently condemned the Lincei dissolved. Cesi's collection of drawings were assembled by Cassiano dal Pozzo (1588–1657) into what he called his 'paper museum' of natural history and antiquities; acquired in the eighteenth century by George III (1738–1820) for the library at Windsor Castle, they are at last being published in a sumptuous multi-volume edition.

The Accademia dei Lincei would be revived in the nineteenth century, but in 1657 Galileo's associates formed, under the patronage of Prince Leopold of Tuscany (1617–75), the Accademia del Cimento, devoted to experiments. Working collectively, and provided with splendid apparatus, they investigated air pressure and other topics ranging from physics to physiology. Then, in 1667, Leopold was made a cardinal and moved to Rome, so the society dissolved after publishing the results of its labours, *Saggi*, a collective work in which no discoveries or experiments were attributed to particular individuals. Soon translated into English, the book played an important part in refining the idea of experiment in science and in increasing understanding of method. As Hooke's friend Richard Waller (c. 1646–1714), the English translator, put it:

> The Experiments are many, and curious, made under the favour of that Prince [Leopold], by the Members of the Academy Del Cimento, men of great ingenuity; and related with much sincerity by the Secretary of that Academy; which Society (I hear) is now scatter'd, and the Hopes of those Benefits the Learned World might justly expect from them, frustrated.[2]

Short-lived societies have thus been important in the history of science, but the brief lifespan of these academies indicates the problem with personal patronage: that, on the death (or change of tack) of its patron, the society collapses, its programme of research is abandoned and its members are dispersed.

In contrast, medical institutions, 'Colleges' of Physicians and of Surgeons run by their members, have a long, continuous history going back to the sixteenth century. But these were notoriously conservative professional associations, essentially serving as white-collar trade unions, rather than being dedicated to the exchange and advancement of knowledge. Professional ethics for them meant behaving towards colleagues like gentlemen. Despite the devastating effects all over Germany of the Thirty Years' War, the oldest surviving scientific academy is rather surprisingly the Leopoldina, now the German Academy of Sciences but in its early days rather different. Founded soon after the Peace of Westphalia in 1652 as the Academia Naturae Curiosorum at Schweinfurt in Bavaria by three physicians, it was essentially a correspondence network, based wherever its president happened to be. In 1670 it began to publish *Ephemerides*, and in 1671 Emperor Leopold (1640–1705) recognised it, giving it a charter to license books and award degrees; in 1687 he renamed it after himself. But it was only from 1878 that it had a base, in Halle, and later still began to hold regular meetings. In the 1660s the most important societies, providing two different models of how to go about things, were those founded in London and in Paris, which had very different political traditions from one another, but also from the loose confederation forming the Habsburg Holy Roman Empire, and from the city-states of Italy and Germany.

When the death of Queen Elizabeth I in 1603 brought James VI of Scotland to the English throne as James I, cross-border raids came to an end but the two nations were not to be fully united politically for a further hundred years. Nevertheless, one can see Great Britain as something like a nation-state, an island with a single sovereign where most people spoke the same language. In continental Europe, with land boundaries and complex tiers of empires, kingdoms, duchies, prince-bishoprics and free cities, only France and the independent federal Netherlands were even remotely like nation-states. The Netherlands was republican, though the prince of

Orange had a weighty position in it; and in the 1650s Britain too was a republic. In both places, elective institutions were potent and successfully challenged the power of their Habsburg and Stuart kings. These were polities in which groups of citizens organised themselves, looked upon governments with suspicion, and resented high taxation and standing armies. Charles II, restored in 1660, prudently never forgot his time as a refugee and exile after his father lost his war with Parliament and was beheaded; less wisely, his brother and successor, James II, tried to seize more power and was ejected in 1688. Britain was a world of checks and balances. In France, it was different: modernity there, after the chaos of the wars of religion, meant absolutism, with the king, and the ministers chosen by him, in control, regulating and deciding everything as far as possible, while the king's magnates, reduced to courtiers, danced attendance at Versailles.

The overturning of ancient authorities that went with the Scientific Revolution led to widespread scepticism, a belief that one could know nothing for certain. If neither Aristotle, Galen and Ptolemy, nor Catholic church teaching could be trusted, there seemed no sure ground for faith. In Greek Antiquity, Pyrrho and those known as the Sophists were sceptics, and so in Rome were Sextus Empiricus (c. AD 160–210) and Carneades (c. 214–129 BC) , after whom Boyle named his spokesman in the dialogues that form *The Sceptical Chymist*. They had alarmed contemporaries because scepticism appeared to open the door to moral relativism, and to undermine religion. Nevertheless, Montaigne, good Catholic and popular mayor of Bordeaux, took as his motto 'Que-sais-je?' – What do I know? – and propagated modern scepticism in his much admired and witty essays. Being closely allied to the 'via negativa' in theology, which proceeds by way of denials – 'God is neither this, nor that, nor that …' – scepticism was by no means anathema to churchmen: doubt might lead to faith. In science, too, Descartes pursued a tactical scepticism in search of certainty that culminated in his 'cogito ergo sum', so that, paradoxically, doubt led to knowledge. Mitigated scepticism, doubting whatever has not been properly tested, is an important part of the scientific attitude. But science is, after all, positive knowledge; if it were all guesswork there would be little point in pursuing it, and the pioneers therefore could not allow scepticism to prevail. They knew they had some right answers, and needed to get together to debate and disseminate them.

In Paris the Minim friar Marin Mersenne published a big book on realism in science, *La Vérité des sciences, contre les septiques* (1625); mathematical works; and an important book, *Harmonie universelle* (1636), on music: but he is especially relevant to our story because of the correspondence network that he maintained with men of science, including Descartes, Galileo, Gassendi, Pascal and Hobbes, and his travels around Europe to meet them. He became a clearing house for scientific information, making discoveries rapidly known and keeping his correspondents up to the mark. Others also took up this role, notably Samuel Hartlib, originally from Prussia though his mother was English, who settled in England about 1628 as war engulfed his homeland. With wide interests in agriculture, Paracelsian medicine, Hermetic philosophy, chemistry, natural philosophy and education, he saw himself as an 'intelligencer', forwarding knowledge through an 'office of address' that he set up, by correspondence and by conversation. He played a part in the political turmoil leading up to the Civil War, was granted a pension by Cromwell, came to know Boyle well and was a neighbour of Pepys. His network or circle, and others, formed an 'invisible college', colleagues who kept in touch by correspondence rather than by regularly meeting together. Hartlib's vast archive survived, is now at the University of Sheffield and has been published in electronic form. Letter-writing, formal and informal, proliferated in seventeenth-century Europe with the growth of literacy and more reliable postal arrangements: in England the Royal Mail was opened up to ordinary users in 1635, with recipients having to pay postage (the envelope and stamp only came two hundred years later). Sending letters further afield did depend upon shipping, embassies or couriers, but also became easier: better communications kept people in contact. Nonetheless, influential though these networks were, there was (and remains) nothing quite as effective as meeting face to face, and actually working in company to promote an end, as a society or less formal group.

Overlapping groups associated with Hartlib's circle, with Wilkins' Wadham College, Oxford, and with the newly founded Gresham College in London discussed natural philosophy and performed experiments during the 1650s. Sir Thomas Gresham, founder of the Royal Exchange (modelled upon that of Antwerp) and an extremely wealthy merchant and financier, had bequeathed his house in the City of London and an endowment to found a college. After the death of his widow in 1596, the lord mayor and City Corporation duly set up a committee to establish it, with courses in Divinity, Astronomy, Geometry, Music, Law, Physic and Rhetoric. The authorities of the universities of Oxford and Cambridge, alarmed at the idea that it might be a rival and grant degrees, were duly consulted and mollified: they nominated the first professors. The idea was that lectures should be delivered in both English and Latin 'in order to render them more extensively useful to all sorts of hearers, whether natives or foreigners';[3] but because John Bull (c. 1563–c. 1622), the first professor of music, could not speak Latin, he was allowed to lecture in English only.

The college became a centre of public lectures given by distinguished men (including, later, Wren and Hooke) and a meeting place, attended by a wide range of people. It was there on 28 November 1660 that a group, meeting after a lecture by Wren and chaired by Wilkins, decided to form themselves into a society for promoting experimental philosophy. They resolved to meet weekly, to exclude all questions of religion and politics, and to seek a royal charter from the recently restored Charles II. This would confer legal status and protection (very important at a time when the scars of civil war, confiscation, exile and military dictatorship were still raw), permanence under a constitution and the right to license books. Following negotiations with the new government by Royalists among the membership, the charter was granted in 1662 (amended in 1663) and the Royal Society began its formal existence. Each year, on St Andrew's Day (30 November), a president, secretary, treasurer and council were chosen, and subscriptions were collected from the Fellows, who had to be properly proposed, seconded and approved. The king did not fund them: royal patronage amounted to little more than benign neglect. They could get on with their own business.

This was to be the model for scientific societies, and indeed much more generally, throughout the English-speaking world. The Royal Society was a club for intellectual and sociable gentlemen, in which the subscriptions of a large number of members, interested in the New Philosophy and its practical possibilities, funded the research and publication of a keen and active few. It met every Wednesday (after the Astronomy lecture) in Gresham College, where it had a meeting room and a repository for books, instruments and rarities until the Fire of London of 1666, when the buildings were requisitioned as a temporary replacement for the burned-down Royal Exchange. It then met instead in Arundel House; it was 1674 before it could move back, after which it continued at Gresham until 1710, when under Newton's presidency it moved westwards, to a house in Crane Court. The Royal Society had, like other novel enterprises, by then had ups and downs: times of considerable activity, and doldrums when Fellows failed to pay subscriptions, discoveries tailed off and the impetus seemed lost. But it had survived.

In 1667 the infant Royal Society licensed its first *History*, by Wilkins' protégé, Thomas Sprat (1635–1713). This work, like Glanvill's writings, was as much an apologia as a history, but it gives a picture of what happened in the early years when Wren was the star performer. The frontispiece in the superior large-paper copies was an engraving featuring instruments and books, a bust of the king being crowned with a wreath flanked by Bacon and Lord William Brouncker (1620–84), the first president, and the Society's arms and sceptical motto, 'Nullius in verba', on top. The motto denotes a determination not to rely upon authority, to rely on nobody's word, but, like della Porta, to examine all claims in the search for truth. Cowley, in his introductory ode, lauded Bacon, who like Moses had spied but not entered the Promised Land, and hailed the Society for carrying his work forward:

> With Courage and Success you the bold work begin;
> Your Cradle has not Idle bin:
> None e're but *Hercules* and you could be
> At five years Age worthy a History.
> And ne're did Fortune better yet
> Th' Historian to the Story fit:
> As you from all Old Errors free
> And purge the Body of Philosophie;
> So from all Modern Folies He
> Has vindicated Eloquence and Wit.[4]

One of the advantages that meetings had over correspondence was that experiments could be newly performed in public. Even nowadays, when we have standardised language and apparatus, repeating an experiment from a published account is not always a straightforward business, and there are at any time published reports that have an uncertain status because what they describe is so far unrepeated, and is perhaps unrepeatable. Newton sent a prism as well as a full description of what he had done to one doubter, so that the latter could do the optical experiment in question exactly as Newton himself had and thus get the same result.

To doubt a gentleman's account of what he had done or seen was to call his honour into question, and gentlemen knew how to resent an injury – with a sword or pistol. In Parliament, members have to be circumspect and are liable to be disciplined by the Speaker: to call another member a liar, for example, is unparliamentary and forbidden, and an instant public apology will be demanded. Even so, eminent MPs fought duels (by then illegal) as late as the early nineteenth century. The Royal Society's 'Nullius in verba', enshrining the basic tenets of the New Philosophy of Galileo, Bacon and Descartes, required tactical scepticism and shared experimental testing of any claim. Performing an experiment, or displaying a rarity, before a gathering of reliable witnesses forming a peer group, having to clarify difficulties and respond to questions in a congenial atmosphere, was and remains a very good way of sharing and advancing knowledge, without calling honour into question. The king gave the Society a mace, to be placed on the table in front of the president, as one is placed in the House of Commons before the Speaker; the meetings thus had a formal character, and enjoyed a parliamentary kind of freedom and equality compared to the outside world, where status was all-important. Despite the Society's commitment to useful knowledge, and interest in the tacit knowledge embodied in craft practices, the Fellows were gentlemen, with income from land or from a profession: when the king told them to elect John Graunt (1620–74), a haberdasher who had published pioneering statistical studies on the London mortality rolls in 1662, they did – and prided themselves on their broadmindedness as they did so.

Men of letters in the early seventeenth century favoured an allusive, witty and flowery style in writing and in rhetoric; but by 1660 a plain style was coming into favour, particularly for sermons by broad-church clergy, or 'latitude men', like Wilkins. Eminent among these was John Tillotson

(1630–94), who became archbishop of Canterbury in 1691 and whose sermons had from the 1660s been hugely admired. He was a great friend of Wilkins, whose stepdaughter he married. Tillotson became a Fellow of the Royal Society in 1672, and the plain workmanlike style he cultivated, supposedly that of artisans rather than wits, had already been adopted as an ideal for scientific purposes. Fortunately, the impersonal tone, passive voice and abstract nouns now favoured (and impressed upon children taking up physics or chemistry) did not come in until the explosive growth of science in the late nineteenth century. The plain style was also characterised as mathematical, with facts and arguments set down logically and dispassionately as in a theorem. In seeking to be full, it could (as with Boyle) become rather rambling and diffuse, but Hooke managed to be pithy and witty enough to be very readable.

Sprat's book contains a report of Hooke's dissection of a dog, and of experiments by Brouncker on the recoil of guns, that would nowadays go into a regularly published scientific journal; but these had not yet been invented. Also to be found in the book, illustrating the importance of practical knowledge and the application of science, are a questionnaire prepared for voyagers, instructions by Hooke on keeping weather reports, accounts of new inventions, and descriptions of the manufacture of saltpetre and gunpowder, the processes of dyeing and the growth of oysters. To keep abreast of work going on at home and overseas, Henry Oldenburg had been appointed as the Society's secretary. Born in Bremen, he first came to England as an envoy in 1653, when he met John Milton, Robert Boyle and members of Wilkins' circle. He became tutor to Boyle's nephew, and from 1657 accompanied him on his Grand Tour, to Paris. There, Habert de Montmor (c.1600–79) was patron of a salon for discussion of science and philosophy and had collected former members of Mersenne's circle; Oldenburg participated and became friends with Parisian savants as the group gelled into an academy with a more formal agenda. Back in London for the Restoration in 1660, he became a founder member of the Royal Society, proving himself an active and able administrator. He maintained a wide international correspondence, notably with Huygens, and in 1665 began publishing the *Philosophical Transactions*, now the longest-running scientific periodical in the world and published by the Royal Society, but at first a private venture building upon a correspondence network and attendance at the Society's meetings. Oldenburg's invention was one of the most important of the whole Scientific Revolution. The papers, unlike those in the *Saggi*, were signed, and in principle were subject to peer review.

Poor Oldenburg was briefly imprisoned in the Tower of London in 1667 because his foreign correspondence caused him to be suspected of spying during the disastrous war with the Dutch. On his release he took up the *Philosophical Transactions* again until his death. At this point there occurred, between volumes twelve (1677) and thirteen (1683), an intermission, during which Hooke published seven parts very like it which he called the *Philosophical Collections*. The *Philosophical Transactions* did not make much money for Oldenburg, and he made his living by translating (especially for Boyle); in 1668 the Society made him a grant of money, which became a regular annual salary of £40 in 1669. The journal was revived and has now been for over two hundred years an august large-quarto publication addressed to experts, with formal papers that mark the completion of a very significant piece of research; but in the early days it was very different. It was then a monthly, in a smaller quarto format, with these 'parts' collected into annual volumes (which began, like books, with an epistle dedicatory). Papers were more like the letters they were evolving from – indeed, often were letters (or extracts from letters with interpolations and cuts by the editor); there were also book reviews, and news items, making the tone informal so that readers would feel part of a group (the secret of any magazine). Oldenburg had made the crucial invention that has characterised modern science: the scientific journal, appearing regularly, securing the priority so important for authors in their research, creating a wide scientific community and keeping it up to date and progressing. Though he, rather than the Royal Society, was the publisher in the early days, the Society provided a ready-made readership of gentlemen – for most of whom science was one interest among others. Since then, journals have proliferated, becoming ever more specialised and less accessible: editors now are usually less personally responsible, and have a team of assistants and referees. They have an editorial board, to lend credibility and/or to assist the editor. Many down the centuries have, like the *Philosophical Transactions* in our time, been produced by a society, but others are the result of publishers being stimulated to fill a niche where researchers lack a suitable outlet or intellectual home.

The editor is an intellectual midwife bringing research to birth, but also a gatekeeper, a builder of reputations who can commission papers, and accept, reject or return for revision and modification whatever gets sent in. It is a position of power that does not ensure popularity. Thus Oldenburg managed to incur the enmity of both Hooke and Newton: Hooke reviewed Newton's optical paper for him and Newton was infuriated by the criticisms; he also later suspected Oldenburg of leaking his unpublished work on the differential calculus to Leibniz. Then, when Huygens announced his improvements to watches, Hooke claimed that the idea was his and that Oldenburg must have prompted Huygens. Editors certainly influence people but may lose friends. They are, however, necessary: by 1683, following Oldenburg's death, the absence of the journal was 'much complained of' and the Society took it on, with the secretary as editor. The preface to volume thirteen (1683) states that 'Although the writing of these *Transactions* is not to be looked upon as the *Business* of the *Royal Society*' they had been so valuable as a register and specimen of work of great variety, and a way of preserving reports of experiments otherwise lost, that the Society had decided to take responsibility for them.

In 1752 a committee was formed to decide thereafter which papers should be published. But the Society remained careful to dissociate itself from the personal views of Fellows, and eschewed a party line. An 'advertisement' published in each volume thereafter stated firmly:

> It is an established rule of the Society, to which they will always adhere, never to give their opinion, as a Body, upon any subject, either of Nature or Art, that comes before them. And therefore the thanks which are frequently proposed from the Chair . . . are to be considered in no other light than as a matter of civility.

Unlike the Cimento, the Royal Society remained individualistic. We may be struck by how long it had taken for Oldenburg's invention to be adopted more than grudgingly as an essential tool for science, bringing honour and prestige to its society or publisher.

Meanwhile in France, the Académie de Montmor (founded 1657) and other circles, and reports about the infant Royal Society, had drawn the attention of Louis XIV's government; and in 1666 the new minister of finance, Jean-Baptiste Colbert (1619–83), as part of his campaign to promote industry and learning, founded the Académie Royale des Sciences. Its particular tasks as a department of state were to map France, improve astronomy and navigation, and ensure that the French remained prominent in science. The arrangements were somewhat casual until 1699, when the king gave it formal status and a constitution, revised in 1716; there was a limited number of salaried members, appointed for life. Aspirants had to await the opportunity to step into dead men's shoes, when the surviving Academicians would propose names to the king. They had a uniform (nowadays, a very fetching one from Napoleon's time) and a place at court, and were expected to assess inventions and innovations, advise when required and play their part as an organ of the state. Unlike members of the Royal Society, they were therefore called upon to express collective opinions; they also offered prizes from time to time for solutions to problems, practical or theoretical, and occasionally sponsored projects. They were the national representatives and exemplars of science, collectively safeguarding orthodoxy and suppressing quackery.

Some Academicians enjoyed a long life, entailing a long wait for their would-be successors: Fontenelle, the permanent secretary, died a mouth short of his hundredth birthday, and we should not imagine that, having survived the perils of childhood, almost everybody died in what we would now consider middle age. Many, especially gentry and professional men, lived on into their seventies and eighties. Retirement and pensions, topics of great interest in our aging society, meant something quite different in the seventeenth and eighteenth centuries. In Johnson's *Dictionary* (1773 edition) 'retirement' means a private abode, private way of life or withdrawal; escaping the hurly-burly of the great city (ancient Rome or modern London) or the court for peaceful reflection and to cultivate one's garden. 'Pension' famously is 'pay given to a state hireling for treason to his country', meaning remittances or bribes paid by the government to pamphleteers and journalists to keep them on side (Johnson was a Tory, and his party out of office). A few such pensions, essentially in the gift of the prime minister, went to worthy and deserving persons (as one had to Hartlib under Cromwell), while kindly or paternalistic grandees might similarly pension off a sick or superannuated old retainer. Otherwise those unable any longer to work could seek outdoor relief from their parishes. Life insurance and annuities came from the Netherlands to Britain towards the end of the seventeenth century, but people did not expect to retire in our sense of giving up their job when they reached a certain age. In the event of serious illness or disability, they might negotiate with a deputy or successor to get a share of their income for the rest of their life. Pepys in his first years at the Navy Board thus had his salary docked to support his predecessor, and the system survived for some posts in the Church of England well into the twentieth century. But essentially people expected to go on working; and Fellows of the Royal Society like Charleton, who found that his medical practice fell off when he was old, might die in poverty.

The French Academicians were salaried for life and so did not risk that fate; but the limit on their number meant that the Académie tended to become a gerontocracy, dominated by, if not thoroughly composed of, the elderly and distinguished. In the early days Huygens was the most eminent member and took the lead; when, in 1672, Louis XIV invaded the Netherlands, he was unhappy but felt that the sciences were above the wars of kings, and stayed put. He subsequently moved back and forth between Paris and The Hague, leaving France finally in 1681; but he felt isolated in his home country and loved the stimulus he found in London and Paris. After the Académie was reformed in 1716, there were forty-four members, in six classes (geometry, astronomy, mechanics, chemistry, anatomy and botany) and three ranks: *pensionnaire, associé* and *adjoint*. In addition, there were foreign members who were invited to join, but remained unsalaried. The secretary had the job of writing *éloges* (obituaries), which were published and became a valuable record and encouragement to others. In addition, from 1666 the Académie published its journal, *Histoire et mémoires*, which became an important vehicle as French science rose to increasing prominence in the early eighteenth century.

This was not the first journal in French concerned with science; the *Journal des sçavans* (later *savants*) had begun publication in Paris in 1665, but, unlike the *Philosophical Transactions* which began a few months later, it did not contain original papers. The editor, Denis de Sallo (1626–69), an associate of Colbert, included book reviews, obituaries, notices of discoveries and inventions in arts and sciences, law reports, university news and mentions of current affairs, sometimes anecdotal. The journal had an uneasy relationship with the censors, and from 1684 an unauthorised edition was published in liberal Amsterdam. It survived right through our period, but became steadily less concerned with what we call science. In 1682 a similar journal, *Acta Eruditorum*, began publication in Leipzig with support from the Elector and under the editorship of Otto Mencke (1644–1707), a professor at the university. This was a monthly, in Latin, and included work done both in the German lands and elsewhere in Europe, thus becoming a valuable if virtual international meeting place. It contained announcements, reviews, abstracts and longer extracts from books and papers, the majority broadly concerned with science and translated into Latin when necessary; a small proportion were original papers. Mencke was succeeded by his son Johann (d. 1732) until, in 1732, a new editor changed the title to *Nova Acta Eruditorum*. Journals publishing essays reviewing recent progress in science, less eclectic than these were, still continue in the sciences, where they play an important role in summarising and assessing what is amusingly called the 'literature' in a particular field; particularly important for graduate students, and for anyone changing tack or working on a frontier between established disciplines. Obituaries too provide an opportunity to stand back and take stock, as well as commemorating (in more expansive style) exemplary lives for the edification and encouragement of the faithful. Published by societies and academies, obituaries and essay reviews helped to build up the group feeling that the great historian Ibn Khaldun (1332–1406) had seen as crucial to any successful movement, giving science a momentum that it had never had before.

Louis XIV was very rich and powerful, and could afford to be lavish. Leibniz urged the king of Prussia to found an academy, and in 1700 he did; but it was underfunded until, in 1744, Frederick the Great (1712–86) reorganised it on a grander scale more closely following the Parisian model. Again, it was international: he appointed Pierre-Louis Moreau de Maupertuis (1658–1759) and Leonhard Euler (1707–83), among others, and the language of the academy, as of his court, was French. Similarly, in Russia, Peter the Great (1672–1725) founded an academy in St Petersburg in 1724 as part of his Westernising and modernising project, but at first there were no Russian members. Under Catherine the Great (1729–96), Euler and other foreigners were brought there, and did distinguished work; the chemist and polymath Mikhail Lomonosov (1711–65), founder of Moscow University, was the first Russian to play a prominent part in it. Though French became the language of the Russian court, the academy's journal was, like *Acta Eruditorum*, in Latin, the learned language in the German lands until the latter part of the eighteenth century; the majority of the academicians were German-speaking, and because graduates everywhere could read Latin this ensured that it was widely read and noted. Other kings, electors and princes founded smaller academies, but they lacked the resources to emulate the French, who by the mid-eighteenth century were in a different league from everyone else, with a critical mass in Paris envied and respected by foreigners – though the Académie Royale des Sciences was beginning to be seen at home, by outsiders and those snubbed by it as an elitist, dogmatic and pompous prop to the *ancien régime*. Britons envied the French their salaries, resources, professional status and respect, and the way a boy in France might aspire to a career in science; while the French envied the independence of the Britons.

In Paris, and then at Greenwich (then just outside London), royal observatories were set up at the same time that the French Académie des Sciences and the Royal Society began their labours. The objective was not simply to increase astronomical knowledge through telescopic observations, but practically to improve navigation. The first astronomer royal, John Flamsteed, had to provide his own instruments. He was a very accurate observer, publishing a catalogue of 2,884 stars; where Tycho with the naked eye had been accurate to about one minute of arc, Flamsteed was usually accurate to ten seconds. As one of the very few professional scientists, actually paid for doing science, and because of his official position, the astronomer royal was an important figure in the British scientific community and in the nation more generally. The observatory slowly developed during the eighteenth century, acquiring the instruments that Flamsteed had hoped in vain to get when the building was first put up. In

Paris they did things differently, and from the beginning the observatory there was better equipped. Its director was Giovanni Domenico Cassini (1625–1712), educated at the Jesuit College in Genoa, and from 1650 professor of astronomy at Bologna. The pope became his patron and, when Cassini was invited to France in 1671, gave him temporary leave to go; but the move proved more permanent, with Cassini becoming a French citizen two years later. He described Saturn's rings, perceived by Galileo as 'ears', as a swarm of tiny satellites; made many planetary observations; and, using data from Guiana and Paris, computed the distance of Mars from the Earth and thus the dimensions of the Solar System. With his son, he began mapping France by triangulation: French cartographers replaced the Dutch as the world leaders in the early eighteenth century. Travellers had found that timekeeping by a pendulum was not exact, because its period varied slightly as one went far north or south from western Europe. This proved that the Earth was not exactly spherical, and Cassini believed that his measurements indicated that it was a prolate spheroid, elongated at the poles, rather than the oblate form favoured by Newton and Huygens, flattened at the poles. In the end, in 1736 Maupertuis took an expedition sponsored by the Académie to the head of the Baltic, where measurements made in the Arctic Circle showed, when compared to those made in France, that Newton had been right. Newtonian physics thus experimentally tested became as much a French as a British science. Observatories became important scientific institutions, centres of activity and meeting places; and in France a dynasty of Cassinis directed the Paris observatory into the nineteenth century.

Another dynasty, founded by Bernard de Jussieu (c.1699–1777) and carried on by his nephew Antoine-Laurent (1748–1836), directed the Jardin du Roi (later des Plantes), the Parisian botanic garden, for well over a century. Botanic gardens, founded as ancillaries to medical investigations, were an indicator of modernity in the seventeenth century, and spread from Italy to France, the Netherlands, Britain and eventually Sweden, where Carl Linnaeus held sway in Uppsala. As centres for teaching and research, they were mostly in universities – Oxford's is the oldest in Britain, founded in 1633 but not appointing its first curator, Jacob Bobart (1599–1680), until 1642. In his first years in post, the Civil War was going on, and when his salary went unpaid Bobart survived by selling produce from the garden. In London, the Society of Apothecaries in 1673 founded its Physic Garden at Chelsea, a spot upstream from London then noteworthy for market gardens, orchards and great houses, so that apprentices could familiarise themselves with useful plants. Later, the eminent physician Hans Sloane (1660–1753) bought the site and leased it to the garden in perpetuity. By 1683 the Chelsea Physic Garden was exchanging plants with its equivalent in Leiden, attached to what was fast becoming the most prominent medical school in Europe, and a major focus of science.

That garden goes back to 1590, when the burghers of Leiden granted the newly founded university a small plot of land, and Carolus Clusius (Charles de l'Écluse, 1526–1609) was appointed to a professorship and put in charge of it. The formal garden which he established, containing a thousand different sorts of plants, has been reconstructed there, as part of the now larger gardens. The Dutch East India Company (VOC) connection brought plants from the Cape and the Far East to the garden; its unique position in Japan, where it retained a toehold in Nagasaki, meant that the garden acquired plants from there that were unknown elsewhere. By the late seventeenth century, it had a greenhouse, and from the 1740s an orangery. Under the polymath Hermann Boerhaave, chemist, physician and botanist, who superintended it from 1709–30 and published a catalogue, it flourished exceedingly and was a magnet for the young Linnaeus, gaining experience and knowledge on his travels to what the Swedes and Britons call the Continent.

Botanic gardens and observatories were attractive places for doctors, naturalists, natural philosophers and mathematicians to meet and keep up to date; places, moreover, where the learned encountered gardeners, lens-grinders and instrument-makers. Meanwhile, museums were evolving from the cabinets of curiosities assembled by powerful rulers such as emperors Rudolph II and Peter the Great and the prince-bishop of Salzburg, the university at Uppsala, and private individuals like Athanasius Kircher, the Danish antiquary Ole Worm and the apothecary Ferrante Imperato (1550–1625) in Naples. The Royal Society had its own such collection, now dispersed, and Grew published a catalogue of it in 1681. Referring to its value in comparative anatomy, he remarked:

After the descriptions; instead of medling with Mystick, Mythologick, or Hieroglyphic matters; or relating stories of Men who were great Riders, or Women that were bold and feared not Horses; as some others have done: I thought it much more proper, To remarque some of the Uses and Reasons of Things. . . . Amongst Medicines, I have thought fit to mention the Virtues of divers Exoticks.[5]

Among those, he included 'Jesuites Bark' and unicorns' horns.

Sloane was secretary of the Royal Society and editor of the *Philosophical Transactions* for twenty years from 1693, and president after Newton from 1727 to 1740; he had become very wealthy, introduced hot chocolate (made with milk) to the British public and was the first doctor to be made a baronet. The enormous collections he had built up, partly by buying other people's cabinets, were bought by the nation to found the British Museum (1753). His cabinet is remembered, and things from his collections are now exhibited, in the 'Enlightenment' gallery in the British Museum. Such public museums, national or regional, with collections including items of natural history, maps, books and instruments, became by the end of the eighteenth century very important institutions, centres for studying natural history and antiquities, and for disseminating knowledge and taste in science and the useful (as well as fine and decorative) arts. That was before anyone thought in terms of 'two cultures'.

Pepys had been responsible for administering the navy, Sloane was a man of parts with powerful friends and patrons, and as president both had been voices for science in government circles; Newton occupied an important position as Master of the Mint. Access to power was what the Society required of its president, rather than especial eminence in any particular field of knowledge; indeed, any such specialist interest might lead him to promote his favourite science at the expense of others. He needed to be a good communicator, and as far as the Royal Society was concerned that meant in English. Universities continued to conduct much of their activity in Latin, and prizes were awarded for verse and prose composed in it; naturalists used a version of it with copious adjectives and few verbs in their descriptions; and gentlemen improved their conversation with Latin tags. But few people actually expected to speak Latin in the ordinary course of life. The same development towards the vernacular took place in France, and French was becoming the language of diplomacy, of elegant courts and indeed also of much science. The decline of Latin meant that international communication was less easy, but contact between university-educated savants and artisans was more straightforward. Those who couldn't read Latin could keep up with progress in science, at least in their own country; and in Britain this was especially important among Dissenters, a group prominent in industry in the eighteenth century, whose education was more vocational than that offered in grammar schools and at Oxford and Cambridge.

Bright lads from the provinces migrated to London or Paris to make their mark in the world of science. Similarly, those from countries on what scholars call the 'periphery' of Europe were drawn towards the centres of activity where they could work with others, causing a steady brain-drain into the metropolitan centres. For the Irish, like Sloane, London was particularly accessible, but Scotland and continental Europe (where Catholics could find co-religionists) were not far away; Nicolas Steno had gone from Denmark to Italy and joined the Cimento; Cassini in Italy and Huygens in the Netherlands were attracted to Paris by the founding of the Académie des Sciences there, and Maupertuis went to Berlin to head Frederick's Königlich Preussische Sozietät der Wissenschaften. Some were asylum-seekers or refugees: Oldenburg came to London to avoid a war; cartographers from Antwerp fled to Amsterdam when the city fell to the Spaniards; and Huguenots like Papin came to England to escape religious persecution in France. Germans and Scots found opportunities in St Petersburg. Some made short visits to foreign countries, on a Grand Tour, to study, or perhaps on embassies. Men of science were a mobile lot. There were national traditions and different ways of making a career in different places; but people moved, bringing knowledge with them, finding more and perhaps taking their new discoveries home.

Travel and expatriate communities were nothing new, but they raised the problem of language in acute form. It is a major accomplishment to be at home in another tongue: words in different languages are never exact synonyms, and translation is not a straightforward business, as scholars had known since the early Renaissance. Sense and style may be in competition: in English, for example, word order is much more important than in more inflected languages, and translations easily lose precision. The very units of weight and length varied from place to place (within and between states), along with technical terms and usages; to aim for plain language was one thing, but metaphor, nuance and allusion would keep slipping in (and are what keep writing lively). As Johann Wolfgang von Goethe (1749–1832) wrote in 1786: 'The idioms of every language are untranslatable, for any word, from the noblest to the coarsest, is related to the unique character, beliefs and way of life of the people who speak it.'⁶ Travellers in Asia, Africa and the Americas reported and struggled with more, and more distinct, languages; but Jesuits in China announced that while there were in that empire several distinct and mutually incomprehensible languages, when the Chinese wrote something down it was intelligible to anyone who could read. Just as English-speakers read 1, 2, 3 as *one, two, three* and the French understand those numerals in the same way but read them as *un, deux, trois* and so on, so different people reading Chinese script would read the same message but in their own tongue. The written characters seemed to represent things rather than words: to be ideograms.

Europeans were not utterly unfamiliar with such things. Renaissance wits had delighted in rebuses, puzzle pictures, emblem books and heraldry. Comenius taught Latin using little pictures of things. The wars of the seventeenth century brought widespread experience of codes and cryptography. In our own time, electronic devices with their icons bring us all into the world of pictograms with its many simplicities and ambiguities. But the news from China was very exciting for those confronted with the new Babel of Europe, and attempts were made to devise some such system that would work universally. Moreover, if words stood for things, then purely verbal disputes would be no more, and obscurantist and flowery rhetoric impossible. In this enterprise, Wilkins went further than most, persuading others to assist him in working out a 'real character' in which facts could be expressed in the Chinese manner. In 1668 he published as a hefty tome his *Essay towards a Real Character and a Philosophical Language*. The book is based upon a supposedly natural classification system reflecting the way things really are, dividing everything into threes and containing some actual transcriptions into the new symbols. Willughby did the zoology and Ray the botany, the latter becoming convinced thereby that such Procrustean simplification was artificial and unworkable, and therefore later returning to Aristotle and developing his natural taxonomy of family groups. Wilkins' magnum opus went the way of artificial languages, into oblivion; but his book is a magnificent monument to an attempt to transcend natural language. It also contains a wonderful illustration of Noah's Ark, with extra sheep to feed to the lions and tigers.

Wilkins was not alone in his endeavours. Francis Lodwick (1619–94), a merchant and Fellow of the Royal Society, also developed an artificial language as part of his quest for concision and accuracy, and for reasonable religion – he recognised that poetry would be impossible in such an impoverished and restricted medium but thought it excellent for communicating facts. In contrast to these arid logical constructs, Wren saw in architecture a rich visual language working across place and time. Describing a classical temple of peace, he wrote:

> No language, no Poetry can so describe Peace, and the Effects of it on Men's Minds, as the Design of this Temple naturally paints it, without any Affectation of the Allegory. . . . [It] is easy of Access and open, carries an humble Front, but embraces wide, is luminous and pleasant, and content with an internal Greatness, despises an invidious appearance all that Heighth it might otherwise justly boast of; but rather fortifying itself on every Side, rests secure on a square and ample Basis.[7]

The temple's architect had displayed 'Wit and Judgement'; but clearly this language expressed mood and emotion, rather than conveying straightforward information – though it might, like icons in the original sense of the term, 'sacred pictures', reveal truths perhaps otherwise inexpressible.

Outside mathematics, signs and symbols did not catch on; science was going to have to use natural languages. As scientific societies promoted what we have come to call a scientific community, those closely involved in it developed jargon, terms more tightly defined than ordinary words and comprehensible across linguistic divides. Well-chosen and definite terms go with the accuracy and precision that Wilkins, Lodwick and others sought. Jargon may thus be necessary and lead to clarity of thought, but can also be a way of excluding outsiders, propping up a blinkered paradigm and expressing commonplaces as if they were profundities. For Samuel Johnson in the mid-eighteenth century, 'jargon' meant 'unintelligible talk; gabble; gibberish', and that is precisely what everyone on the outside feels. The philosopher and physicist John Herschel (1792–1871) wrestled with the difficulty of finding words to get science across when the plain language of artisans seemed to fail. On the one hand, he wrote: 'There is scarcely any person of good ordinary understanding . . . who may not be readily made to comprehend at least the general train of reasoning by which any of the great truths of physics are deduced, and the essential bearings and connections of the several parts of natural philosophy.' On the other hand, going deeper, where ordinary language was no good, entailed

> A sound and sufficient knowledge of mathematics, the great instrument of all exact inquiry, without which no man can ever make such advances in this or any other of the higher departments of science, as can entitle him to form an independent opinion, on any subject of discussion within their range.[8]

For Herschel, we might note, 'man' covered both genders, for he was among those who much admired the writings of Mary Somerville (1780–1872). Anyway, by the later eighteenth century it was clear that, as Galileo had declared, the universal language in which the Book of Nature was written was not hieroglyphics but mathematics.

While parts of science were for experts, right through the Scientific Revolution societies were amateur and much scientific writing remained accessible to any educated person, and indeed to craftsmen. Following the founding of scientific societies and academies, science became a part of the culture of eighteenth-century Europe and its colonies in the Americas. And as it became accepted as an honourable and perhaps useful vocation, the status of medicine, the learned profession depending upon scientific knowledge, steadily rose. There were great advances in anatomical knowledge, though these made little difference in actual treatments, and there were new remedies, herbal and mineral. The development of statistics made possible informed concern with public health: for the seventeenth century was an age of plague, though there was little that doctors could do about it, and of endemic diseases. It is to medicine that we now turn.

CHAPTER 8

LIFE IS SHORT, SCIENCE LONG
THE HEALING ART

ASTRONOMY, WITH ITS BASIS in awe and wonder, its connection with navigation, and the way it could be elucidated by mathematicians, was in Antiquity and in the Scientific Revolution an exemplary science. Optics and mechanics seemed to be going the same way. But while astronomers could make predictions, it was impossible to experiment on stars and planets, and dazzling counterintuitive theories about the Earth's being in motion were far from the understanding and concerns of the mass of mankind. They worried about their health and their future, and looked to the heavens for signs and portents. What, after all, might stars be *for*? It was odd that there were so many: navigation would have been easier with fewer. The general view was that the heavens and the Earth were intimately connected, so that events here could be forecast by the stargazer. Comets, meteors and eclipses were significant for individuals as well as nations: Romeo and Juliet were star-crossed lovers. Our temperaments depended upon the zodiacal sign and disposition of the planets when we were born. Astrology was a necessary part of medical knowledge, a key to individuality. William Lilly (1602–1681) in his *Christian Astrology* (1647) gave instructions for advising clients about health and sickness, as well as marriage, career, contracts or travels. Like other men of science, he was confident in asserting his independence of authority:

Perhaps some will accuse me for dissenting from Ptolomey [*sic*]; I confesse I have done so, and that I am not the first, nor shall I that have done so, be the last; for I am more led by reason and experience, then by the single authority of any one man, &c.[1]

Astrology brought astronomy down to earth, and almanacs were an immensely popular kind of publication, consistent bestsellers year after year: in Restoration Britain between 300,000 and 400,000 are reckoned to have been sold annually, meaning one between every three households. In 1696 William Daniel, MP for Marlborough, bought *Mercurius Anglicanus* by George Parker (1654–1743), a 'student in Physick and Astrology', and generally used it as a diary, filling it with dinner engagements, and births, deaths and marriages among his friends, but on 2 August he noted a portent, that the Sun 'appeared of a very pale colour and weak light'.[2] Each month there is a table of saints' days, ides and nones from the Roman calendar, sunrise and sunset, phases of the Moon, high tides in London, and 'observations', such as that in August, through Saturn, 'the Fair Sex are much menaced with divers Calamities, – Abortions and other sorrows for pregnant Women &c'. At the back are an explanation and defence of astrology, and a number of advertisements (beyond which, some frugal person a hundred years later used the blank pages to record some experiments on dyes). Astrology was normal science, and thus valuable to medical men as well as diarists for the light it cast on patients' constitutions.

So was the experimental science of chymistry, important through pharmacy. Since Paracelsus, the latter had included mineral as well as animal and vegetable drugs, which gradually found their way, despite conservative opposition, into the pharmacopoeia. The wonderful thirteenth-century manuscript herbal *Medicina Antiqua* includes pictures, done at two different periods, of doctors at work, administering, for example, a potion made from a plant.[3] Many medicinal plants are illustrated along with the doctors or midwives using them. This is charming rather than useful, because the illustrations are too stylised for easy recognition; and in the Scientific Revolution, the increasingly naturalistic pictures of plants in printed books were generally separated from those illustrating anatomy and surgery. John Parkinson (1567–1650), apothecary and botanist to James I and Charles I, choosing a

Latin title for his beautifully illustrated book on garden plants, punned on his own name: *Paradisi in Sole* (Park-in-Sun). Since 'out of the ground made the Lord God to grow every tree that is pleasant to the sight, and good for food', describing Paradise/Eden as a park is apt, and the pun works.[4] In the frontispiece we see the Garden, tended by Adam and Eve (whose nakedness is discreetly veiled by hair and vegetation) under the all-seeing eye of God, and prominent in the middle distance is the 'vegetable lamb' of Tartary, growing on a stalk and cropping the grass within its reach. Unfortunately we do not find within the book directions for cultivating one ourselves.

In botanic gardens and on field trips such as those recorded by the apothecary Thomas Johnson (c.1600–44), apprentices and students learned about the herbs and 'simples' from which most potions and ointments were made; and new arrivals like tobacco, potatoes and chocolate found their place in medical treatments. Empirical knowledge gradually supplemented, and then supplanted, the 'doctrine of signatures' – the idea that God gave us hints from the appearance of plants (heart-shaped leaves, perhaps) what disease they might relieve. To increase their usefulness, Parkinson's book and the *Herball* of John Gerard (1545–1612), its largely plagiarised 1597 edition much improved by Johnson in 1633 to compete with Parkinson, had after their Latin and English indexes a 'Table of Vertues'. In Gerard's the reader could see how 'to represse overmuch *Vomiting* of Choler', 'to stay *Wambling* of womens stomacks being with child' or 'to provoke *Bodily* lust'.[5] Medical problems do not change very much over the centuries. Two novelties did, however, bring relief to our ancestors: from the East came opium for the relief of pain, its addictive properties and side effects not yet comprehended, and still in crude form in which controlled dosage would be hard to determine; and from Peru in the West 'Jesuit's bark'. This product of the cinchona tree is the source for our quinine: it is a specific for malaria, the ague then common in western Europe ('Agues', 'Quotidian', 'Quartan', 'Tertian' and 'Shakings' have many entries in the *Herball*), and with its bitter taste fulfilled the expectation that nasty medicine is good for you. But the bark was even trickier to administer than opium because the strength depended on exactly which kind of tree it had come from, and at what season; and apothecaries were not always scrupulous, surreptitiously substituting a cheaper alternative for some hard-to-get constituent (as must always have happened with unicorn's horn, in demand as an aphrodisiac, or for deterring spiders). So, again, determining dosage, or even deciding if a treatment worked, was hard. There were, however, so many remedies listed in the *Herball* for every human condition that it is amazing everybody was not cured.

They weren't. Cities were both magnets and deathtraps. Not until 1874 could the registrar general report with satisfaction that for the first time in its history London was self-supporting in terms of population, as with a new sewage system its death rate fell beneath its birth rate – though even then 30 per cent of babies died before they were three years old. Other cities were just as unhealthy, their populations maintained and increasing through immigration from the healthier countryside. Particularly in the seventeenth century, on top of the general run of endemic diseases came terrifying epidemics or visitations of bubonic plague. Neither prayer nor medicine alleviated them: nobody knew how they spread. In 1636 in Newcastle the scrivener Ralph Tailor (1611–69) was kept busy drawing up last-minute wills for plague sufferers; the social fabric did not collapse despite the horrific death rate. But generally those who could afford to do so, fled – sometimes unwittingly carrying plague-infected fleas in their baggage. In Britain the final visitation was the Great Plague of 1665 which caused Newton to leave Cambridge for Woolsthorpe. This is also the best documented because we have the diary kept by Pepys, who remained in London throughout. In 1722 Daniel Defoe (c.1660–1731), author of *Robinson Crusoe* and *Moll Flanders*, projector, spy, businessman, pamphleteer and agitator, published *A Journal of the Plague Year*, a novel made topical by an outbreak in Marseille. Defoe used contemporary records and memories in working up what reads like a vivid personal account of the horror of the plague in London, with its quarantining, pest houses, mass graves and dead carts, 'searchers' determining causes of death, and the practical and moral questions which the epidemic raised.

Plague also lay behind *Natural and Political Observations* (1662) by John Graunt, a study of the London *Bills of Mortality*, which were first published during a plague outbreak in 1592, and from 1603 were printed regularly each week – available at a cost of four shillings a year to subscribers. The searchers were 'ancient Matrons, sworn to their Office',[6] and Graunt tabulated their reports and in a pioneering statistical investigation saw what could be learned from them. His chief objective was to investigate the size and growth of London's population, at which he made an educated guess; but more interesting to us are the conclusions about diseases and public health that he arrived at on the way.

Causes of death recorded by the searchers were very various. In 1665, 68,596 people died of plague (out of 97,306 deaths in toto). Over the twenty previous years 6,384 had died of plague, while 44,487 had died from consumption and coughs, 23,784 of apoplexy or 'suddenly', 15,759 from agues and fevers, 10,576 from smallpox and 9,623 from dropsy. Little children were hit particularly hard: 32,106 of the dead were (unweaned) 'Chrisoms and Infants', 14,236 died from 'teeth and worms' and many more would be in other categories. Graunt reckoned that about 36 per cent of children died before the age of six, and 60 per cent before they were sixteen. Noting that some diseases declined and others rose, he inferred that this indicated different diagnoses of the same disease: thus deaths from what was called 'Tyssick' fell but those from 'Cough' rose in proportion. He also found that, during plague epidemics, deaths from some other diseases increased; here, he believed that searchers had been induced to record plague deaths wrongly, so that families could avoid being quarantined. After epidemics he reckoned that immigration brought the population back to its previous level within two years: London and other cities remained hugely attractive to prodigal sons and daughters keen to get on. Things were very different in colonial America, where the native inhabitants died from European diseases but the colonists' big and healthy families meant that the population doubled each twenty-five years. On 23 June 1659 the poet Anne Bradstreet (1612–72) wrote of her eight children, all of whom survived, as the youngest took flight:

> I had eight birds hatched in one nest,
> Four cocks there were, and hens the rest.
> I nursed them up with pain and care,
> Nor cost, nor labour did I spare,
> Till at the last they took their wing,
> Mounted the trees, and learned to sing.[7]

Graunt's patron, whose influence in his work was strong and to whom it was sometimes attributed, was William Petty (1623–87). The son of a draper, he had gone to sea, landed up in Normandy in 1637 and attended the Jesuit College in Caen; he returned thence to England, but on the outbreak of the Civil War went to Leiden to study medicine, and on to Paris where he met Hobbes, Gassendi and Mersenne. In 1646, back in Oxford, Petty completed his medical studies and in 1651, in good odour with the new regime, became professor of anatomy. As he began to dissect Anne Greene, hanged in Oxford, his actions instead resuscitated her; in consequence he achieved great fame, fellowship of the College of Physicians and a readership at Gresham College. In 1652 he was appointed physician to Cromwell's army in Ireland, and in 1654 his proposals to use teams of soldiers to map the conquered lands for redistribution were accepted. The results of this 'down survey' did not please everyone, but they enriched Petty, who acquired enormous estates. He became a member of the English and Irish parliaments, was a founder of the Royal Society and, having made peace with the restored Stuart government, was knighted in 1661. He was first president of the Dublin Philosophical Society, founded in 1661, and designed a catamaran, which unfortunately capsized in a great storm. His maps of Ireland were published in 1685, and he collected enormous quantities of data about Ireland and England for analysis, some of it published in his posthumous *Political Arithmetic* (1690). His son became Lord Shelburne (1737–1805), and another descendant became prime minister and patron of Joseph Priestley.

Petty's life showed how far a medical man with energy, charm and an eye on the main chance might get; and how important statistics were becoming to rulers grasping just how little they knew about the countries and people they governed. This opened up useful territory for applied mathematicians. Analysis of risk was essential for the modern banking and financial services that began in the Netherlands and came to London at the end of Petty's lifetime; and in particular vital statistics were crucial for life insurance and annuities. Huygens had in 1657 supplemented such empirical, inductive statistics by deductive analyses of games of chance. British baptism and funeral records were too patchy for accurate computations, so Halley in his actuarial calculations used more reliable data from Breslau for

1687–91, hoping they were comparable. This pioneering work was taken up by Abraham de Moivre in his *Doctrine of Chances* (1711, 1738 and 1756). Another exile, a French Protestant educated in Saumur, he fled the country when the Edict of Nantes was revoked in 1685, and in Britain became a protégé of Halley and Newton. He derived better formulae for insurance calculations, anticipated the law of large numbers, and was an important figure in the wider paradoxical effort to find the laws of chance: but the great age of 'lies, damned lies, and statistics' would be the nineteenth century.

Some who were trained in medicine went far in different fields, like Petty and the philosopher John Locke. Thomas Secker (1693–1768), who was raised a Dissenter, studied medicine in England, Paris and Leiden (MD 1721), was later ordained and became a distinguished archbishop of Canterbury – probably the only one to have dissected a corpse. But for most such students medicine remained their vocation, and a striking feature of the times is the increasing importance of medical men. Paying for a medical visit augmented or superseded the traditional remedy of prayer as doctors took over roles that had previously been played by the clergy, especially in relation to what came to be seen as mental illness; they began to enter and dominate what had been the female sphere of childbirth, and used their training to good effect in chemical and biological sciences. Despite the lists of remedies in herbals, and their clinical experience, there was little that they could do, apart from encouraging the patient with their bedside manner – and perhaps making things worse by bleeding, purging and shutting windows. Alcohol and opium as painkillers, and 'Peruvian Bark' for fevers, were standby remedies; during the later eighteenth century Francis Home (1719–1813) in an Edinburgh hospital, and other medical men in the army and navy, and in private practice, began to assess the effects of treatments more systematically, but it was not straightforward. No doubt doctors' treatments, based upon experience, knowledge of their patients and their families, and the healing power of nature, often worked: we know how effective placebos and the nostrums of alternative medicine can be even in these days of evidence-based medicine. In Petty's time and thereafter, a cheerful, easy and encouraging manner, and breezy self-confidence, were essential; and the family's doctor might hope to become their friend.

Indeed, though sometimes mocked, as by Gassendi's pupil Molière (Jean Baptiste Poquelin, 1622–73) in his plays *Le Médecin malgré lui* (1666) and *Le Malade imaginaire* (1673), doctors seem nevertheless to have been generally trusted and often loved as avuncular sources of wise advice, as Charles Darwin's grandfather, Erasmus (1731–1802), evidently was.

Ever since Hippocrates, an important aspect of medicine had been prognosis: patients needed to know how things would turn out, and particularly whether they were going to die. Deathbeds were important: a good death, reconciled to God and neighbours, rounded off and made sense of a life. Medical men became much more involved in the care of the dying across our period, and doctors' bills began to show up in executors' accounts. Hippocrates, many of whose case studies end with the death of the patient, had famously described the indications of imminent death in the changes to the face; good prognoses would depend upon further close clinical observation of disease in patients with various constitutions and illnesses. Thomas Sydenham (1624–89), called the English Hippocrates, was a puritan who left Oxford when the Civil War broke out in order to join the parliamentary army. He saw heavy fighting and became a firm believer in Providence, which had spared him. Returning to Oxford, he helped in the purge of Royalist 'malignants' there; but by 1655 he was practising medicine in London, and living near Boyle's sister, Lady Ranelagh (1615–91). He recommended a new and mild regime for smallpox; rejected the heavy bleeding and purging in vogue for other diseases; was caustic about fellow physicians who fled the Great Plague; and, believing medicine was still short on evidence, made numerous careful case studies (following Hippocrates' example) in the 1660s and 1670s, published as *Observationes* in 1676. He combined his puritan faith in Providence with seeking laws and investigating constitutions in a framework of natural religion. But by 1700, instead of trusting to Providence, the prudent were taking out insurance.

Sydenham, with degrees from Oxford and later Cambridge, was made a licentiate of the Royal College of Physicians in 1660 at the restoration of Charles II, and was among the elite in his profession. Beginning in Salerno, and then in Bologna, Padua, Montpellier and other universities all over Europe, would-be physicians had since the Renaissance studied anatomy, *materia medica* (originally herbs, but later mineral substances also) and regimen, with Latin as their language and Greek and Roman texts as their guide; they had little hands-on experience. The learned nature of their profession gave them the status of gentlemen, which in turn meant that they were not supposed to do manual work: their diagnoses were based upon external symptoms, pulse-taking and examination of urine. Their visits were expensive, and fashionable doctors became wealthy: most people never consulted a physician, but in the German lands towns appointed and housed salaried physicians to look after the health of citizens. Further down the medical hierarchy came the apothecaries, shopkeepers who learned their trade through apprenticeship, made up physicians' prescriptions and sold drugs – but who were not supposed to charge for advice. In London, their company (or guild) separated from the grocers in 1617; its members became in effect general practitioners and community pharmacists, and as a result for the next two hundred years they were engaged in disputes with the physicians about their right to practise. To combat fraud, at the Apothecaries' Hall in London drugs were prepared and tested for purity.

Surgeons were manual workers, craftsmen who also learned their trade by apprenticeship, and in London they separated from the barbers (with whom they had been formally joined in 1540) only in 1745. They were licensed by a bishop and treated the wealthy under the supervision of a physician; for ordinary people, they were another kind of general practitioner, and again their activities might bring them into demarcation disputes with physicians. In France, Ambroise Paré (1510–90), after a spell with a barber-surgeon in the provinces, came to Paris and worked at the Hôtel-Dieu hospital; he then joined the army, where he observed that the current practice of cauterising wounds with boiling oil made things worse, and inaugurated a milder regime. He began using ligatures to reduce the loss of blood in amputations, soon acquiring a high reputation for skill based on direct observation and experience, and served as surgeon to four kings of France. He was called upon to treat wounds caused by guns, which he perceived as detestable agents of death and destruction:

> These horrible monsters of Canons, double Canons, Bastards, Musquits, field pieces; hence these cruell and furious beasts, Culverines, Serpentines,

Basilisques, Sackers, Falcons, Falconets, and divers other names not onely drawne from their figure and making, but also from the effects of their cruelty. . . . Wee might clearely discerne, that these engines were made for no other purpose, nor with other intent, but onely to be imployed for the speedy and cruell slaughter of men.[8]

Back in civilian life, he published an autobiography, as well as a treatise on surgery: the latter aroused the ire of the physicians, who believed that only they should publish medical books. Moreover, the book was in French because Paré knew no Latin, and therefore was too accessible. The book's merits were such that the attempt to suppress it was vain.

Surgeons could not read the Latin publications by Vesalius and other learned authors; they relied chiefly upon experience, like apothecaries, and manual skills rather than formal learning. Very few made their way into the upper reaches of society, and it was only in the eighteenth century, when the university medical schools of Leiden and then Edinburgh began encouraging apothecaries and surgeons to take courses alongside physicians, that they began to move up in the world. In the later eighteenth century surgeons in Britain's Royal Navy ceased to be petty officers, messing with the boatswain and carpenter, and joined the officers (the captain ate on his own). The Leiden graduates who founded the Edinburgh medical school in the 1720s brought with them a vision of a much more practical course of training for physicians than had previously been the custom. In the new-style medical school, the professor came to be seen as a specialist in a branch of medicine, who delivered formal lectures following a published syllabus that students would buy and annotate, and who might, from their notes and his own, work up and publish a textbook. Previously, Scottish students had been taught everything by a single 'regent master', in what was therefore more like an apprenticeship. Now both being up to date and having hands-on experience began to matter. All this led, despite the medical hierarchy, to a further blurring of distinctions, which had always been hard for the physicians to enforce – especially when a 'quack' found a powerful patron. Medicine in eighteenth-century Britain was in effect a free market.

The most dramatic feature of Renaissance medical schools was the annual public dissections of the corpse of a murderer. Artists and civic dignitaries as well as medical students came to watch. Because it took some days, to keep the smell of decay within bounds the event happened at the coldest time of the year – it thus almost formed part of the carnival celebrations before the penitential season of Lent. Like a play in the round, the action took place in a cylindrical theatre, with spectators standing closely packed in an array of galleries. Traditionally, the professor read from Galen's text, and the demonstrator performed the actual dissection. There must have been times when the text and the demonstration were at odds, but every human body is a bit different and, given how few dissections anyone would have witnessed, any discrepancies could be ascribed to individual variation. But Vesalius, getting out of his chair at Padua and doing the dissecting himself, was prepared to take issue with Galen over the heart, showing that the septum dividing it is not permeable: Galen had supposed that it filtered dark red venous blood, turning it into bright red arterial blood. This was a tremendous challenge to one of the most respected medical authorities, the author of the largest corpus of ancient Greek writing that we have. Vesalius' published lectures gained credibility through their superb engravings in classical style (with rural backgrounds) in which a human body is gradually stripped down to a skeleton. In the frontispiece Vesalius himself is shown getting down to work on a corpse surrounded by students and spectators. Reading Vesalius' text and looking at the magnificent pictures made one a virtual witness to his dissections. Bishop John Cosin (1595–1672), an editor of the Church of England's 1662 Book of Common Prayer, owned a copy; to learned men in an unspecialised age, anatomy was an appropriate subject whatever their profession.

For anatomists, Vesalius' challenge raised new questions. If blood could not get through the septum, there must be another way of getting at least some of it from one side of the heart to the other, and turning it red and frothy. Miguel Servetus (1511–53), a heretic executed on Calvin's orders, had got an inkling of the alternative route, through the lungs and back to the heart (the lesser circulation or pulmonary transit); but his books were burned with him and it was Matteo Colombo (1515–59), Vesalius' deputy, critic and successor, who demonstrated it through dissections and experiments on animals. Thus modified, Galen's view survived. The venous and arterial systems were distinct: blood, which he believed was made in the liver, carried nourishment in the veins to the limbs, while the bright red arterial blood was the source of life, further filtered in the brain into finer form to flow through the nerves. It seemed to most people that there was no longer anything to puzzle over; but to the curious, careful observation can always reveal fresh problems. William Harvey studied medicine at Padua (after Cambridge), in 1600–2, and on his return to England began a very successful career as a physician: appointed to St Bartholomew's Hospital, he was nominated in 1615 and 1618 as Lumleian Lecturer. Temperamentally conservative, he nevertheless went beyond Colombo with his theory that all blood circulated. Arterial blood flowed outwards from the heart to the extremities in ever-narrowing vessels invisibly joined to the veins that brought it back to the heart, whence it went through the lungs (where it was turned bright red) and completed its circle (or, strictly, figure of eight).

For Harvey, the most important point was that this theory made sense of the valves in the heart which maintain this one-way traffic, and the valves in the veins that allow flow only towards the heart. Galen (and, recently, Colombo) had performed dissections and vivisections, using apes and pigs because they resembled humans. Harvey looked instead to animals very different from us, notably fish, which have no lungs and therefore no pulmonary transit, and whose hearts are simpler than ours. Using comparative anatomy to illuminate the human body was not new, but Harvey in his experimental and observational argument used it from right across the animal kingdom, to which we humans after all belong. Initially, he had been puzzled to note that, contrary to what Galen said, the pulse (or a spurt of blood from a punctured artery) accompanied the contraction, or systole, of the heart, squeezing the blood out. Where did all the blood forced out at each beat go? It couldn't be constantly used and replenished, but must go round and round, becoming darker in colour on the way. By 1618 he was sure about the circulation, driven by that powerful contraction; but he was uneasy, as a traditionalist, an admirer of Aristotle's teleological or functional approach that had set him thinking about valves, and he realised that his views would be controversial.

Harvey's little book *De Motu Cordis* (On the Motion of the Heart, 1628) featured only one illustration, showing the valves in the veins (which had previously been known but not understood). It was set out like a disputation, building up a cumulative argument. We appreciate most easily the chapter in which he argues quantitatively that the blood is pumped through the system much too fast for it all to be used up and renewed, so it must go round and round:

> In the course of half an hour, the heart will have made more than one thousand beats, in some as many as two, three and even four thousand. Multiplying the number of drachms propelled by the number of pulses, we shall have either one thousand half ounces, or one thousand times three drachms, or a like proportional quantity of blood, according to the amount which we assume as propelled with each stroke of the heart . . . a larger quantity in every case than is contained in the whole body![9]

Neither Harvey himself nor his contemporaries were greatly impressed with this argument, depending as it did on reasonable assumptions about how much blood the heart ejected at each systole: they had anyway believed that diastole, or expansion, was the heart's power-stroke, and critics were unconvinced by Harvey's account of that. They were schooled to respect book learning and ancient knowledge, clinical experience and teleological reasoning, and to suspect novelty as quackery and mathematics as superficial. But, in the event, the support (in general if not in detail) of Descartes, a mechanist and innovator very different from Harvey, proved very important; as is usual with successful scientific innovation, the rising generation went for the new idea. Not long after Harvey's death, microscopists duly observed the blood moving through the capillaries he had hypothesised. We ordinary mortals are taught to be deeply sceptical about guesswork like that in science, but just as the microscope vindicated Harvey, so (much later) the telescope vindicated Copernicus' hypothesis that stellar parallax caused by the Earth's motion would be observed: scientific geniuses and madmen defy the ordinary rules.

Critics pointed out that believing in the circulation had no clinical consequences. It might have weakened belief in blood-letting, but it didn't: Harvey himself on his deathbed asked to be bled (from his tongue). In the short run, the anatomical advances from Vesalius to Harvey made no more difference to medical practice than the contemporary Copernican system made to navigation. Harvey said that his practice had fallen off after he published his theory, but research did bring patronage. Vesalius attended the emperor Charles V, the most powerful man in the world; Colombo was called by Pope Pius IV to Rome; and Harvey became physician to King Charles I, to whom he had dedicated his book. The flowery preface compares the role of the heart in the body to that of the king in his kingdom:

> The heart of all animals is the foundation of their life, the sovereign of everything within them, the sun of their microcosm, that upon which all growth depends, from which all power proceeds. The King, in like manner, is the foundation of his kingdom, the sun of the world around him, the heart of the republic, the fountain whence all power, all grace doth flow.[10]

This must have pleased Charles as, frustrated by Parliament, he entered upon his period of personal rule. Having to dance attendance at courts is not necessarily compatible with research, and in Harvey's case he found himself also backing the losing side in the Civil War. But he had the opportunity to dissect deer from the king's park in pursuit of his other Aristotelian interest, generation and embryology, on which he published his other very important book, *De Generatione Animalium* (The Generation of Animals, 1651), with a preface outlining his philosophy of science. He was convinced that all living creatures came out of an egg, and that both parents contributed equally to the development of new and unique offspring, in a process called epigenesis: rival theories, called preformation, had it that coition called to life seeds stored in the testicles or the ovaries of one parent. Just as his theory of circulation depended upon invisible capillaries carrying blood from arteries to veins, so here he inferred the existence of mammalian eggs, unobserved for two more centuries. Again, this had no short- or medium-term implications for medical practice.

Physicians were more interested in patients than diseases, and in prevention rather than cure; at court, the doctor was there to ward off illnesses that he had small hope of curing, by promoting a healthy lifestyle appropriate to the constitutions of his charges, on whom he kept a careful eye. In general, the inverse care law applied: most medical attention always goes to those who least need it. But we are all mortal, and when things went wrong with his wealthy clients, the physician would do his best and call in colleagues for their opinions. He might call in a surgeon. What he would not do was to send his patients to hospital. People expected to be treated at home, even when major surgery like amputation was required, a grim business undertaken as a last resort. Pepys, who had had a bladder stone removed, thought the outcome so satisfactory (though it may be why he had no children) that he celebrated by inviting the surgeon each year to a party. Many other patients were not so fortunate. They would be nursed by the family, a term that then included household servants. Care would be augmented when necessary by other women who would watch with, tend or nurse the patient, generally relying upon experience rather than training. If the patient died, the same women would probably 'lay out' the body and help at the funeral and the wake. Most causes of death were not seen as threatening to those who looked after the sick, and the fees paid to such women were low.

However, in the case of desperate epidemic and contagious diseases (smallpox and especially plague), doctors were reluctant to visit, and often fled. Attendants and nurses might look after those infected, but for higher-than-usual fees. Then, because of the danger to the public, the authorities in towns made regulations, enforced by a pest master and wardens, governing the quarantining of victims' families by shutting up their houses, and charging them for these services. Whereas the sick could usually determine their own treatment, sufferers from plague were passive: they could not choose their attendants, and might indeed be conveyed willy-nilly to a pest house to die or recover there. If they pulled through, when the risk of infection was past a surgeon might dress their plague sores. But pest houses, like mass graves or 'plague pits', were a response to a crisis, very different from what happened ordinarily.

Monasteries and nunneries normally had infirmaries as part of their buildings, and monks and nuns when ill would be treated there, with herbal remedies and a diet less austere than that ordinarily available to healthy members of their community (especially on fast days). They would expect to be nursed by their fellows, in due course to die there, and to be buried within their cloisters. They might treat others in the vicinity who were sick – as might parish clergy, as the only learned people in the neighbourhood, in the way that missionaries did in the twentieth century. But hospitals were for the poor, and were usually not primarily for acute diseases, accidents and emergencies: they were religious foundations, saving souls rather than lives, for the most part more like almshouses or our care homes and hospices than modern hospitals, as indeed the word's connection with 'hospitality' indicates. A few such foundations survive. 'Christ's Hospital' in London was (and, on a new site in Sussex, is) a charitable boarding school for poor but able boys. While the deserving poor or aged might hope to get a place in a hospital, often on the nomination of a patron or subscriber, and remain there for life, a few hospitals did treat and discharge the sick; by the early eighteenth century secular hospitals more like today's were being established, supported by government on the Continent and in towns in Britain by subscribers (who had rights of nomination) and by collections in churches on 'Spittal Sundays'. Physicians sought honorary positions there, going in perhaps one day a week as a highly respected charitable duty, and gaining experience with patients who had to obey doctor's orders, while surgeons practised there, and trained (as 'dressers') their apprentices. These hospitals might be associated with dispensaries, providing outpatient treatment; and in contrast to the medieval ones did not allow patients to stay indefinitely, discharging 'incurables'.

Whereas most of our ancestors through the ages must have stunk, cleanliness came to be seen as next to godliness by the early eighteenth century, and ancient Roman practices like frequent bathing were revived. The Romans had particularly valued mineral springs like those in Bath, and now water cures came back into fashion. These might involve sea-bathing on breezy northern beaches. People swam naked in rivers, but in seaside Scarborough bathing machines were in use by 1736: little huts on wheels in which people could change into their ample bathing costumes, they were wheeled out into the waves so that their occupants could descend modestly into the cool grey water of the North Sea. The fashion spread gradually as the century wore on and sea-bathing around the coasts of Europe began to be perceived as enjoyable rather than just chilly and medicinal.

Most people took their water cures inland, in resorts like the eponymously named Spa in Belgium, Aachen in Germany, Carlsbad (Karlovy

Vary) in Bohemia and Bath in England, where treatment could be combined with sociability and pleasure. The patients would take a communal bath, usually before breakfast, adjourn to the pump room to drink the spring water (tasting strongly of its minerals), then promenade before dinner (steadily getting later in the afternoon) and spend the evening sociably. The dandy 'Beau' Nash (1674–1761), expelled from Oxford, briefly an army officer and a law student, was in 1705 appointed Master of Ceremonies at Bath, where he presided over public balls conducted with great splendour. Called the 'King of Bath', he enforced a regime that depended not on rank but on easy and enjoyable mingling and conversation, adherence to a dress code and etiquette, and firm regulation of hours: even princesses had to stop dancing an hour before midnight. Bath was developed into the elegant city we see now, financed by subscriptions and gambling profits, and in the season attracted between eight and twelve thousand visitors.

Many of the latter must genuinely have been sick rather than bent on pleasure, as we can see from the numerous monuments to the dead in Bath Abbey. William Oliver (1695–1764), a Cornish doctor, educated in Cambridge and Leiden and elected FRS in 1730, collected subscriptions from wealthy patients and friends to build a Water and General Hospital in Bath, where in 1740 he was appointed physician. A major supporter was his philanthropic friend Ralph Allen (1693–1764), another Cornishman who, backed by the eminent soldier George Wade (1673–1748), had secured an extremely lucrative contract to run mail-coach services. He promoted a scheme for making the Avon navigable to transport stone and goods to Bath, where he became mayor in 1742. There he built a handsome mansion, Prior Park, on top of a hill, up which a railway line was constructed to carry materials: a technical feat that attracted much attention. Oliver remains famous for inventing a savoury biscuit, the Bath Oliver, still made today, rather than for any great medical success. But water cures, much less draconian than the bleeding, purging and cauterising otherwise enforced, achieved a long-enduring popularity.

Tunbridge Wells and Harrogate in England were long established as spas and, though without quite the éclat of Bath, became fashionable and elegant; and other towns with mineral springs hoped to join in the boom. Physicians could help: they were called upon to analyse the water and point to any constituents that might relieve specific complaints. Chemical taxonomy was in flux, but chemists were agreed about the important distinctions between acids and alkalis, metals and non-metals. The classical world's fascination with the sympathies and antipathies among things was transmuted with a metaphor borrowed from marriages into the notion of 'elective affinities': gravity was universal but, like humans, chemical substances sometimes loved each other and formed lasting combinations, and sometimes didn't. After they had reacted together, the addition of a more attractive and active substance would destabilise the relationship. Chemists were beginning to tabulate these affinities, and to sort out various kinds of 'earths', classifying them so that they could be identified systematically. In the field, especially in Sweden with its iron industry, rapid qualitative analysis of ores by fusing them on a charcoal block using a blowpipe was practised. Taste, smell, texture and colour were important criteria, learned by experience. Like qualitative analysis of organic substances, quantitative analysis, especially of waters (where the mineral content was small), remained tricky and the results questionable; but medical men, using the knowledge they had picked up in their training, seemed the appropriate people to do it – again, expanding their sphere of activity.

As mentioned above, medical men found further opportunities in laying claim to the right to treat those whom we, following them, call the mentally ill. In the country, those with learning difficulties or mental handicap were in medieval times looked after in their families and communities, no doubt with variable kindness and attention; there were things they could usefully do, and they might be thought of as holy fools. The insane were looked after in the same way, but counselling them was the particular responsibility of the clergy. Doctors perceived them, like other patients, as suffering from an imbalance of their four 'humours': blood, phlegm, yellow (choler) and black bile. 'Melancholy', indicating an excess of black bile, was fashionable in the early seventeenth century, notably in the Netherlands and in Jacobean England: it went with being intellectual, wearing dark colours, perhaps toying with a skull, and admiring 'Vanity' paintings that indicated the passage of time and the futility of worldly ambition. The formidably learned Robert Burton, who spent all his adult life at Oxford, overcame what we would probably call depression by writing about it, adopting the persona of Democritus junior, because Democritus, as well as being an atomist, was famous for laughing at the follies of mankind. Burton's great book *The Anatomy of Melancholy* (1628) is a wonderful source of anecdote, wisdom and curious knowledge. His 'remedies against discontent', based upon authorities ancient and modern, often came down to 'count your blessings' and think of the misery of 'many myriads of poor slaves, captives, of such as work day and night in coal-pits, tin-mines, with sore toil to maintain a poor living'.[11] But he did discuss medicines too, noting reports of the 'admirable and profitable' but dangerous antimony, and 'Tobacco, divine, rare, superexcellent Tobacco, which goes far beyond all their panaceas, potable gold, and philosopher's stones, a sovereign remedy to all diseases' – but only when used in moderation and under medical advice. The object of medical treatments of melancholy was to purge the excess of black bile and restore the patient's equilibrium: purges, vomits and bloodletting were the unpleasant staples of all contemporary doctoring.

The melancholy might respond to cheering up or to a purge, but some sufferers were violent or became so, and in cities full of strangers there was little community support. By the seventeenth century the insane were being confined. In Paris the Salpêtrière was converted from a gunpowder manufactory into a hospital, used as a dumping ground for the feckless poor, the insane and prostitutes. There is a long and continuing history of imprisoning the mentally ill like criminals. In 1654 Louis XIV ordered a new building, including a handsome and very large chapel, but it was hated as a kind of gaol and stormed in the Revolution. In a suburb of Paris there was another hospital, the Bicêtre, opened in 1642, projected for old soldiers but in the event chiefly used as an orphanage, where the insane and prostitutes were also confined. In London one of the medieval hospitals, Bethlehem or Bedlam, was for the insane; 'Bedlamite' came to mean lunatic. Bedlam was destroyed in the Great Fire of 1666; rebuilt to a handsome design by Hooke, it became a perverse kind of tourist attraction, where chained lunatics were viewed with sorrowful wonder or as dreadful examples of degeneracy, as in the famous series of paintings and engravings by William Hogarth (1697–1764), *The Rake's Progress* (1733–5).

It was accepted that suffering was a consequence of someone's moral failure, probably but not necessarily the sufferer's. Self-examination and confession were therefore the first steps when the clergy were confronted with the insane. They knew that God moved in a mysterious way, and might as with Job allow Satan to test His faithful servants (illnesses of all kinds were a test of character). But we are all sinners; and Jesus, after forgiving sinners, cast out the demons that had been troubling them. As in first-century Palestine, the Devil walked abroad seeking whom he might devour, and awareness of his omnipresence and activity was awakened in the Reformation of the sixteenth century. The ritual practices and folk religion by which his wiles, and the spells cast by witches under his aegis, were combated, by exorcisms, spells or water from holy wells, were denounced by Protestants and, following the Council of Trent, by Roman Catholics too as superstitious practices. Ghosts and spirits were identified with demons, and the machinations of witches and necromancers became much more frightening. The Scientific Revolution coincided with the peak of witch-hunting: witchfinders interrogated suspects, and William Harvey was called upon to inspect a woman to see if she had the extra nipples that witches required to nourish their 'familiar', such as a cat – she had not. In both Protestant and Catholic countries, 'cunning men' and 'wise women' were regarded with a mixture of awe and deep mistrust; and aggrieved, alienated, usually impoverished and female, social outcasts found themselves at the focus of terrible and possibly fatal hostility.

Against this background, Simon Forman (1552–1611) taught himself astrology and began to practise medicine. Imprisoned in 1579 and 1587 and on other occasions too for black magic but undeterred, in 1588 he began calling up angels and spirits. He managed to read very widely, and was contemptuous of orthodox Galenic medicine for its neglect of astrology and alchemical remedies. He treated himself for plague, lancing his sores and taking a potion, and then did very well financially by treating other sufferers. A notorious and Faustian figure, pursued by the physicians as a quack, Forman was involved in sexual scandals, accused of supplying love potions and poisons to some highly placed ladies, and of necromancy. Nonetheless, he remained in high repute as a magus among astrologers. His pupil, who inherited his papers, was Richard Napier (1559–1634), a very different sort of man but with similar interests, whose sixty volumes of neatly kept records survived and have been brought to light by Michael Macdonald in his *Mystical Bedlam* (1981).

Unlike the plebeian and self-educated Forman, Napier went to Oxford and became a Fellow of Exeter College, studying theology. In 1590 he was ordained and became rector of Great Linford in Buckinghamshire, where, hating preaching (and employing curates to do it for him), he practised medicine from 1597 until his death. He had fitted in studies with Forman and knew Dee; remaining an orthodox Anglican, he combined magic, religion and the new science in his treatments. Whereas physicians spent much time visiting patients, Napier's came to him in his rectory, and he might see up to fifteen in a day. They came from all classes, and some from a great distance. What he did was to cast a horoscope for them and enquire about their symptoms. He would then give them a purge, perhaps also a vomit; order blood-letting; and say a prayer with them. He might then consult the archangel Raphael for a second opinion. While this was anathema to some of his puritan fellow clergy, respect for his deep and wide learning and for his huge theological library, and success with his patients, kept him out of trouble. Only a small proportion of his patients were mentally disturbed, and he sought to discriminate between mental affliction (some had troubles enough to make anyone desperate) and the work of witches and demons. Clearly, being treated by a sympathetic and understanding magus was very different from being chained in a dungeon. Our ancestors did not distinguish diseases as assiduously as we do, except for plague and smallpox, but anyway illness inevitably has two aspects: the objective, 'What have I got, and what's the prognosis?', and the subjective, 'Why is this happening to me, and how should I interpret its message, and make the best of it?' It was the same in Napier's day, and while we might call the first medical and the second spiritual or existential, a good doctor still helps with both questions.

Faced with insanity, puritans would resort enthusiastically to prayer to cast out devils; they distrusted astrology for its fatalism, while any contact with supposed angels to them seemed diabolical. Catholics in contrast might go for exorcism with bell, book and candle. To middle-of-the-road Anglicans, all these things were repugnant, and by the later seventeenth century they were content to see insanity as a medical rather than spiritual problem. Then, if the purge and vomit failed to work, restraint was one of the options, and at the end of our period the straitjacket was being used even for King George III. Private madhouses were set up to care for dotty, awkward or tiresome relatives of the better-off under medical supervision. Whether all this was to the advantage of the sufferers is by no means clear: a talking cure from a sympathetic clergyman like Napier, especially if he was a good listener, must have been much more congenial. Doctors might also become involved in legal cases: for instance, where a patient was being made a ward of court for protection against fraud and rapacity or where someone accused of a crime pleaded not guilty by reason of insanity. The ill-starred, the moonstruck, the lunatics, the crazy, the possessed and the melancholy became the mentally ill. It was one of the most striking developments of the Scientific Revolution.

As noted earlier, the female world of childbirth provided further opportunities for intrusion by medical men, and the outstanding figure in this field was William Smellie (1697–1763). An apothecary-surgeon, he practised in Lanark in Scotland from 1720, becoming interested in obstet-rics. In 1733 he became a member of the Faculty of Physicians and Surgeons of Glasgow, and by 1739 was in London, where he established a pharmacy and from 1741 began teaching midwives, who had previously always learned on the job and been licensed, and medical students. He devised models for teaching, designed forceps, improved methods for delivering breech babies and attended poor women free on condition that his students could accompany him. In 1745 Smellie obtained a Glasgow MD; by the time he died, 'man-midwives' had become respected as obstetricians and had seized leadership from women. Among his pupils was William Hunter (1718–83), who became an extremely successful obstetrician and teacher in London, running a private medical school; his classic book *The Gravid Uterus* (1774) was superbly illustrated with engravings by Jan van Rymsdyk (c. 1730–88). Hunter had obtained the fresh corpse of a woman who had died late in pregnancy during a cold winter and dissected it carefully in the chilly days that followed. His brother and protégé, John (1728–93), also a surgeon, became the greatest physiologist and comparative anatomist of the day. No doubt many pregnant women benefited from their expert help, but it is worth remembering that there have always been people who suffer as a result of scientific and technical advance. In this case it was midwives, who now found themselves de-skilled and their experience, knowledge and techniques disregarded or appropriated, often to general disadvantage. We have benefited enormously from science, but its progress has left behind its casualties and its blind spots.

A medical education could be the gateway to the sciences more generally, opening other careers. Our ancestors expected (as we do) that their doctors would be primarily concerned with keeping them healthy and fighting disease. To do that they needed to learn *materia medica* and anatomy, as well as, by the eighteenth century, physiology, concerned with functions and processes in the manner of Harvey, and anatomy, concerned with structure. Microscopes became increasingly important, though they improved little optically until the nineteenth century. Discoveries in anatomy and physiology brought intellectual excitement as real as that in astronomy, but might have little impact upon clinical activity: there (as in other technologies), the improvement of techniques and instruments (such as forceps), and the publication of results of best practice using them, were much more important. In teaching, the production of the detailed wax models that can still be seen in medical museums allowed students to familiarise themselves with anatomy; and William Hunter, by injecting wax during a dissection, proved that the foetal and maternal bloodstreams are distinct. Though astronomy had gone from medical courses by 1700, university students now learned chemistry, comparative anatomy and botany. That meant that a degree in medicine was not only a route into a profession, but also the best way to get into parts of chemistry, zoology and botany that had little direct connection with human health.

Thus Boerhaave, who studied at Leiden and then took his MD at Harderwijk, became in 1709 professor of botany at Leiden, where he much improved the botanic garden. In 1714 he was made professor of practical medicine, and he greatly promoted bedside teaching: students would accompany him each week on his round of twelve selected patients. In 1716 Peter the Great was among his pupils, and more professional students from all over Europe were attracted to Boerhaave's Leiden, the world's leading medical school. In 1718 he became professor of chemistry, and was soon recognised as the great pundit in that field too: in 1724 a 'pirated' edition of his highly successful and respected chemical lectures, based upon notes taken down by students, was published without his permission. Not until 1732 did he bring out an authorised version, by which time the pirated edition had found translators into English and other languages; the official one joined it on the market, and in one form or another the book became the standard compendium throughout Europe for a generation. Boerhaave as a chemist did not make any great original discovery, but his lectures and publications were clear, straightforward and empirical, and as a teacher he acquired an immense reputation, giving him an important place in the history of that science.

The kind of artificial, convenient classification based on external characters that botanists worked out for plants was applied in medical education in 'nosology', the science of diseases grouped not by causes, as Boerhaave had hoped to do, but by symptoms. Going back to Sydenham's idea that diseases were real entities rather than imbalances of an individual's humours, François Boissier de Sauvages (1706–67), professor at Montpellier, director of the botanic garden there and friend of Linnaeus, devised a taxonomy. In the Edinburgh medical school William Cullen (1710–90) made it the basis of his extremely influential teaching. He believed that a 'natural' system based on causes was hypothetical and confusing to students, whereas grouping diseases by symptoms into class, order, genus and species made diagnosis (the word was also used by botanists) straightforward. Nowadays, though plants and many illnesses can be classified more 'naturally' using chemical and other methods, identifying mental illnesses still has to be done Cullen's way.

Medicine with its background in sciences is a kind of technology, in which knowledge brings power to improve human life. In other fields too, the Scientific Revolution similarly involved educated and inquisitive people engaging in the improvement of crafts, requiring new instruments and apparatus from skilled artisans, and promoting inventions; while, like the surgeons bringing hands-on experience to physicians, craftsmen taught their social superiors valuable lessons. It is to this world of crafts and inventions that we turn next.

CHAPTER 9

MAKING THINGS BETTER
PRACTICAL SCIENCE

A S SIR JOSEPH BANKS later remarked of one of the first examples of laboratory science applied to the saving of lives, Davy's safety lamp, such discoveries 'Cannot fail to Recommend the discoverer to much Public Gratitude, & place the Royal Society in a more Popular Point of view than all the abstruse discoveries beyond the understanding of unlearned People Could do'.[1] Intellectual excitement is thrilling and admirable, but science is and always was valued and supported chiefly because, as with medicine, we hope with Bacon that it will bring us lives that are longer, safer and more comfortable as well as more interesting. Crafts and inventions go back to the beginnings of human history; but while grandees and mandarins in ancient, medieval and Renaissance society admired and coveted handsome buildings, tapestries, pictures, jewels and accoutrements, manuscripts beautifully illuminated, and clocks, they did not as a rule concern themselves overmuch with how they were made. Such banal practicalities were unsuitable for ladies and gentlemen. Rather, they competed for the services of skilled practitioners; and their competitive and aesthetic interests extended inevitably to fortifications, ships and guns, bringing patronage to engineers, shipwrights and artillerymen who developed expertise in these fields. Trades became 'mysteries', entered by apprenticeship, controlled by guilds, and largely based upon tacit and oral knowledge. With the coming of printing, texts such as those by Agricola and Biringuccio on mining and

Brunschwig on distillation brought technical practice to the notice of the learned, and by the sixteenth century machines were beginning to exert their fascination upon the growing group of readers (perhaps especially male ones).

The new curiosity among the learned about the natural world, ancient traditions and secret writings, and the workings of mills, printing presses and clocks, made the knowledge embedded in craft mysteries worth investigating. Getting one's hands dirty in the field or the laboratory, contriving and manipulating apparatus, experimenting, became things that gentlemen should not despise. Know-how was valuable. This practical turn towards sooty empirics, working by rule of thumb, deplorable though it seemed to the fastidious and intellectual Hobbes, was as important in the Scientific Revolution as any changes in natural philosophy. Social distinctions certainly did not fade away; but just as the painter in Renaissance Italy had ceased to be an artisan and become an artist who consorted with gentry, so tradesmen like Graunt and lens-grinders like Leeuwenhoek became part of the genteel Royal Society along with the better-born Wilkins and Boyle. Conversely, Newton was not the only young gentleman who made mechanical models in his spare time.

Early connections between natural philosophy and technology began with a practical invention or contrivance; then later, perhaps much later, came the inquisitive man of science keen to know how and why it worked. Empirics came first, in technology as in medicine. Babylonians made calendars before anybody knew about planetary orbits; spectacles antedated Kepler's optics; gunners' experience of ranging preceded the dynamics of Tartaglia, Galileo and Newton; and the hands-on knowledge of potters, miners, dyers, apothecaries and stained-glass-makers long anticipated testable chemical theory. Before 1800 it is hard to find examples of what we are taught to expect: technology as applied science, following a few steps behind scientific understanding gained in the laboratory, in the field or in armchair calculation. Things were generally the other way round: the working of machinery, perhaps already being improved by craftsmen making a series of small changes or adjustments, was investigated by a savant and understood in more 'philosophical' terms. In the light of this, it might be redesigned, made to work much better or its scope extended. Thus the working and limitations of long-familiar water pumps were initially a puzzle: from the consequent experiments and thinking on the atmosphere that began with Torricelli, the fruit was Boyle and Hooke's air pump and Papin's pressure cooker, the 'digester'. Similarly, spyglasses achieved by trial and error were transformed into telescopes both by means of better techniques of lens-grinding and by taking into account the laws of reflection and refraction found by Kepler, Snell, Descartes and Newton (whose specimen reflecting telescope intrigued the Royal Society). Science grew out of crafts.

Exemplifying this process, Bernard Palissy (c.1510–90), having trained as a stained-glass maker and finding his trade in decline, introduced into the ancient craft of pottery the kind of systematic experimentation that must have happened long before in China:

> More than twenty-five years ago I was shown an earthen cup, turned and enamelled with such beauty that I was immediately perplexed. . . . I came to think that if I were able to make good enamels I could make pottery vessels and other things of good design, for God had given me the knack of knowing something about drawing; and immediately, without thinking that I had no knowledge of clayey earths, I started to look for enamels like a man who gropes in the dark. . . . I crushed all sorts of things that I thought could be used. . . . I would buy a number of earthen pots, and after breaking them to pieces, I would put the things I had crushed on them, and after marking them I would write down the drugs I had put on each one, as a reminder; and after I had built a kiln to my liking, I put those pieces in to bake.[2]

Building on his successes, he made pots and naturalistic rustic figurines that sold well, and grottoes for the patrons he attracted. An entrepreneur and man of parts, he carried his anti-authoritarian, experimental interests into other arts and sciences, publishing two books set out as discussions between 'Theory' and 'Practice'. Like Paré the surgeon, he was a Huguenot saved from the massacres in 1588 by the queen mother; but two years later he was thrown into the Bastille, where he died.

So we shall not find any simple story of one-way traffic between science and technology; though, in the wake of Bacon, we shall find plenty of interactions, and 'projectors' arising to promise, boost or claim new inventions. English examples were the wealthy and quixotic Edward Somerset, 2nd marquis of Worcester (1601–67), courtier and inventor of an unsuccessful steam pump; and Samuel Morland (1625–95), a would-be courtier widely despised as two-faced in those interesting times, who was also involved with water-works, a clumsy calculating machine and a speaking trumpet. More soberly, in Britain as in France trades were systematically investigated to find the laws that lay behind rules of thumb developed by artisans. There were notable advances in clocks and watches, instrument-making, air pumps and early steam engines, water wheels, fortifications, roads and bridges, optical devices, printing and engraving, mapping and the making of globes, and agriculture.

The ancients had water clocks, much improved in Islam; and mechanical clocks were in use long before the Scientific Revolution. They kept poor time by later standards, and therefore needed resetting from sundials when the sun shone, but they became increasingly splendid, decorative and complicated – huge in cathedrals, bejewelled and elaborate in courts. Germany became a particular centre of clockmaking, where, as well as utilitarian models, handsome clocks were made to be admired and as presents to be taken on embassies, notably to Turkey. By 1390 the finest clocks like that at Wells Cathedral had two hands, for hours and minutes (not straightforward to read), and might well have other dials showing the month, sign of the zodiac, phase of the Moon and other information. In 1582 Galileo had noticed, it is said from watching candelabra in Pisa cathedral, that pendulums of the same length, whatever (within limits) the amplitude of their swing, take equal times for each beat. He hoped this property might be used in regulating clocks, but it was not until 1657 that Huygens devised a way of actually doing it. By 1600 watches were also available, and Malvolio, the puritan steward in Shakespeare's *Twelfth Night* (c.1601–2), in reverie imagines himself a count: 'I frown the while; and perchance wind up my watch, or play with my – some rich jewel.'[3] In 1665 Pepys, a rising civil servant in the Navy Office, was given one by a client; they were status symbols, sometimes jewelled and elaborate, but kept poor time. Our ancestors could not be punctual until after 1675, when Hooke and Huygens invented the balance spring (and squabbled over priority). Clockwork was also adapted by ingenious workmen to drive automata, celestial globes, armillary spheres and orreries.

One of the most famous clockmakers was Thomas Tompion, a blacksmith's son, who in 1671 was admitted to the Clockmakers' Company of London. A versatile craftsman, in that year he also cast a church bell of 4 cwt (200 kg) for St Lawrence's church, Willington, in Bedfordshire; shortly afterwards he came to the attention of Sir Jonas Moore (1617–79), a mathematician and surveyor (notably of the Fens of East Anglia where large-scale draining was going on), who in 1669 was made surveyor general of the Ordnance, based in the Tower of London. He was the major promoter of Greenwich Observatory, and of Flamsteed; in 1674 he commissioned a clock for the Tower and a long-case (grandfather) clock for himself from Tompion, and introduced him to Hooke, with whom he worked on watches. Also in 1674 Tompion had made a quadrant for the Royal Society, and in due course he made another for Flamsteed as well as clocks for Greenwich. By 1677 he was also advertising mercury barometers from the larger premises to which he had moved in 1676. In 1675 he had made a balance-spring watch for King Charles II, and further royal patronage followed; he invented a 'repeater' clock that would strike the last hour again when a cord was pulled – for example, in the middle of the night; and by the 1680s his workshop was making well over a hundred clocks and fifteen to twenty watches a year. He built up his workshop like a Renaissance artist's studio, developing a house style, mechanisation and some division of labour. Maintaining high standards of workmanship, he trained over a hundred craftsmen in his lifetime; his timepieces were in demand all over Europe, and were bought for courts or as presents to take on embassies. Master of the Clockmakers' Company in 1703, he died a wealthy man, and was buried in Westminster Abbey: a long way for a blacksmith's son to have come. Technical skill, as the Scientific Revolution gained momentum, could lead with suitable patronage to great social advancement.

Tycho had made his instruments as good as he could, estimating error and well aware of the limits of accuracy within which he worked. Increasingly, measurement and precision became a feature of the Scientific Revolution, demanded by observers and experimentalists, and more widely in a society becoming more punctual and conscious of time and distance.

The urge for precision bore fruit in clocks regulated by pendulums and watches by balance springs, incorporating minute hands and even seconds hands; accurately engraved scales on instruments; better optical glass; fine balances for assayers; and attempts to impose uniformity in place of local standards of distance, weight and other measures. The clocks for Greenwich were important because they made it possible to measure what Newton called 'absolute, true, and mathematical time',[4] which flows equably (and on which 'Greenwich Mean Time' is based), in contrast to the 'common time' from sundials, which is dependent on the Earth's irregular motion round the Sun. Moreover, pendulum clocks like Tompion's meant that time could be measured more accurately than angles: without the need to swivel, telescopes could therefore be firmly and securely mounted and the exact times noted at which stars or planets traversed the crosswires set up in the eyepiece. Using them, Flamsteed made the observations that for Newton and Halley superseded Tycho's and confirmed their views about planets and comets. Halley and others recognised that accurate clocks also provided a way of determining longitude: the time – for instance, noon – changes by one hour for every fifteen degrees of longitude, as we know very well from air travel. Pendulum clocks could be erected in observatories anywhere and set to local time at noon using a sundial, and longitude then found by observing an eclipse, and comparing the local time with that at Greenwich or Paris, where the same phenomenon was recorded. But at sea, in pitching and tossing sailing ships, such clocks were not practicable.

Tompion had successors, instrument-makers who not only assisted the virtuosi but joined them and founded enduring businesses, or dynasties. George Graham (c.1673–1751) came from a humble background in the north to London to serve as an apprentice. There, as a journeyman he subsequently worked and lived with Tompion, married his niece and in 1711 became a partner in his business, inheriting it in 1713. He continued to make improvements to clockwork. About 1720 he introduced the dead-beat escapement, and in 1721 the compensation pendulum, in his very accurate 'regulator' clocks. Pendulum rods expanded with heat, affecting the running of the clock; but by using a glass jar of mercury (which would expand upwards) as the 'bob' of the pendulum, Graham neutralised this effect. Master of the Clockmakers' Company in 1722, he had in the previous year been elected FRS; he served on the Royal Society's Council, and presented twenty papers at its meetings. As well as clocks, he made orreries and astronomical instruments, including one taken on the Paris Académie des Sciences' expedition to Lapland in 1736 which demonstrated the flattening of the Earth at the poles. He too was buried in Westminster Abbey; and his pupils, notably Thomas Mudge (1715/16–94), helped to maintain the reputation of London as a great centre of clock- and precision instrument-making.

Meanwhile, much earlier the wealthy Scottish laird John Napier, who had studied overseas, busied himself with projects for improving fertilisers, the effectiveness of artillery and chariots. Strongly Protestant, in 1593 he published a commentary, dedicated to James VI, on the New Testament Book of Revelation or Apocalypse, portraying the pope as Antichrist. In what might seem to us (but probably did not to Napier's contemporaries) a surprising change of tack, he then, in 1614, published in Latin his book on logarithms, a novel mathematical technique for transforming multiplication into addition and division into subtraction, translated into English in 1616. Henry Briggs (1561–1631), a Cambridge mathematician and puritan, excited by reading about logarithms, hastened to Scotland to see Napier, and in 1615 lectured about this new technique at Gresham College, where he was a professor. In 1617 he published logarithms more conveniently based upon 10, the form in which they remained an essential part of the intellectual equipment of mathematicians and scientists up to the two last decades of the twentieth century.

Napier had himself made logarithms easier to use for practical purposes with his famous 'bones', a kind of abacus with numbers engraved upon strips of bone or ivory with which multiplication and division could be done by juxtaposition. William Oughtred (1575–1660), another Cambridge mathematician and clergyman whose patron was Thomas Howard, 15th earl of Arundel (1608–52), collaborated with Elias Allen (1588–1653), an instrument-maker whose workshop was near Arundel House in London, in devising a more convenient version. In 1622 they devised 'circles of proportion' which could be moved rather like volvelles to perform calculations, and shortly afterwards straight versions, rulers in which a slider engraved with numbers moved against a logarithmic scale. As the 'slide rule' ('slipstick' in the USA), this valuable tool for rapid calculation was essential equipment for students of physics and chemistry as well as for every engineer until the electronic calculator was invented. Briggs' textbook *Clavis* (1631, 3rd edn 1652) was used by Seth Ward (1617–89) and by Wren. A mathematical innovation by a practical-minded gentleman was rapidly taken up by astronomers, notably Kepler, and others; and an instrument, incorporating it and simple to use, was devised when a mathematician and an instrument-maker worked together.

The mathematician and instrument-maker might be one person. Measuring angles, the altitude of stars or of the Sun at sea, or between landmarks in surveying on land, was difficult even with quadrants like Tompion's: you have to squint along the arms, first one way and then the other. John Hadley, eldest son of a wealthy country gentleman and thus at home in the Royal Society of which he was a Fellow (1717) and vice-president from 1728, built an improved model of Newton's reflecting telescope in 1722. Then, in 1730, he devised an octant with mirrors so arranged that the images coincided and the angle could be easily read off: it was confusingly called Hadley's Quadrant because it covered ninety degrees. In 1732 he added a spirit level so that it could be used even when the horizon was obscured. By 1754 Christ's Hospital Mathematical School, training navigators, had ordered that its boys be equipped with this device that had eclipsed all others.

In 1757 John Campbell (1720–90), sailor son of a Church of Scotland minister, suggested enlarging it to measure up to 120 degrees, and making it in brass rather than wood; and in 1759 the instrument-maker John Bird (1709–76) made one, of 8-inch (20 cm) radius: this was the sextant, and soon became a standard piece of equipment. Campbell had sailed round the world with Anson, becoming FRS in 1764 and an admiral in 1778. He had used a quadrant to measure the angular distances between the Moon and stars as a possible method for navigators to find their position, working with James Bradley (1692–1762), Savilian professor of astronomy at Oxford and then, from 1742, astronomer royal in succession to Halley. Bradley had a passionate interest in instruments, and had demonstrated the bulging of the Earth at the equator by sending one of Graham's clocks to be set up in Jamaica, where because of the gravitational difference it duly ran slow. A prediction Newton had made could be verified after his death with a precision instrument, rather as Harvey's had been with a microscope. Such triumphant vindications of theory are a feature of the progress of science, brought about through a scientific community and the availability of talented craftsmen associated with it. And the making and trading of precision instruments, sometimes gleaming in the polished mahogany, silver, steel and brass that made them desirable for demonstrations and make them covetable and collectible today, became important and profitable, bringing prestige and wealth to the successors of Tompion and Graham, with London a great centre of activity.

The story of clockwork shows how close collaboration between craftsmen who had served an apprenticeship to their trade, navigators and gentlemen with a university education advanced both what we could call pure science (Bradley discovered the aberration of light as well as elucidating the Earth's shape) and technology in the form of accurate grandfather clocks and of instruments leading to the sextant. There is a similar story behind the machines that supplied power on a scale unattainable by human or animal labour. Windmills and water wheels go back to ancient times and were widespread by the sixteenth century when Dutch polders and then English fens were drained with pumps powered by the wind. By then the classic western water wheel, mounted on a horizontal axis, was also being used for grinding and to work trip hammers for fulling and other processes connected with the cloth trade. But life is not all work, and in Italian Renaissance palazzi the gardens were enriched with fountains and cascades for which the water might be pumped up by a wheel. The fashion spread to Salzburg, where the prince-bishop had a water garden with fountains and grottoes in which unsuspecting visitors might be sprinkled; and also to France.

In 1684 Louis XIV inaugurated the vast system of waterworks at Marly, where 'la Machine' pumped water 500 feet (150 m) from the Seine at a rate of up to a million gallons (4.5 million litres) per day to supply the water for Versailles, especially its spectacular fountains. The chief engineer (a job Morland had sought) was Arnold de Ville (1653–1722), for whom the king had a château built nearby; there were fourteen wheels of about 38 feet (11 m) diameter, and a complex system of rods, pumps and reservoirs going up the hillside. It was one of the wonders of the age, and the noise below and the fountains above must have been stunning.

The Huguenots Salomon de Caus (1576–1626), who had worked for Louis XIII (1601–43), and his brother Isaac (1590–1648) brought French landscaping to Britain, and worked with the architect Inigo Jones (1573–1642) in introducing Renaissance style. The great country houses of the nobility and gentry came to include water features, but water wheels were also applied to practical purposes. Thus there was a great wheel under an arch of the old London Bridge, pumping water from the Thames for domestic and industrial purposes: the many arches of the bridge blocked the incoming tide, so that the water was fresh, if not clean. That wheel, and most everywhere else, was undershot: that is, its base dipped into the flowing water, which turned it anticlockwise. By the eighteenth century, engineers, notably John Smeaton (1724–92), who won the Royal Society's Copley Medal for the work, demonstrated the greater efficiency of breast wheels, where the water struck the wheel on the level of the axle, again driving it anticlockwise; and, better still, of overshot wheels, where the water comes over the top of the wheel, driving it clockwise. Both these kinds use the weight as well as the flow of the water, and require buckets rather than just vanes to hold it. The flow of water and its eddying had fascinated Leonardo, and stimulated Newton to devise hydrodynamics in refuting Descartes' theory of vortices; and in the nineteenth century Sadi Carnot (1796–1832), founder of thermodynamics, was inspired in his theory of the flow of heat through a steam engine by the water wheels that his engineer (and French revolutionary) father Lazare (1753–1823) had studied. Smeaton was more practical: among his other works were a new 32-foot (9.8 m) wheel for London Bridge in 1768, and water-powered blowing engines for Carron Iron's blast furnaces in Scotland; the company's works adjoined the Forth and Clyde canal for which he was chief engineer.

Smeaton and other engineers designed and improved steam engines, stimulating Carnot to wonder if there was a limit to their efficiency. Although the Industrial Revolution in its earlier stages depended upon water power, the steam engine became the iconic source of energy that transformed first Britain and then other European countries and the USA.

It was the first device in which heat produced by chemical reaction generated work, marking a radically new departure in engineering. Ultimately, it also brought into being the energy-based 'classical physics' of the later nineteenth century, for, as its great exponent William Thomson (Lord Kelvin, 1824–1907) put it, natural philosophy owed more to the steam engine than the steam engine owed to it. Historians have puzzled over just how much science did go into the steam engine, and where the components assembled into the first engines had their origin. Popular credit for the steam engine goes to James Watt (1736–1819), who, after abortive trials at the Carron ironworks, entered into partnership with Matthew Boulton (1728–1809) from 1774 to make his improved and much more efficient machine. But by then steam (or, strictly, atmospheric) engines were already commonplace in the north of England and in Cornwall, and their origins and development come firmly within the time of the Scientific Revolution.

As mines for coal and tin went deeper, they ran into water. In a drift mine, dug into the side of a hill, this could be drained away with an adit emerging lower down; but deep mines entailed pumps, and they required power. Horses might be used, going round and round in a gin gang, but two- or four-horse power with lengthy and heavy arrangements of rods going down a mine was soon insufficient; and there might be no convenient stream for a waterwheel. In a booklet of 1659–60, R. D'Acres (probably a pseudonym) described raising water using the elements of air, water and, more hypothetically, fire: he did not illustrate his machine, in which hot air from a furnace was admitted to a vessel 'all overhead in a *Pond* or *Cistern*',[5] which, being then cooled, would suck up the water. Nothing came of that, but in 1698 the military engineer Thomas Savery (c.1650–1715) patented a device which he illustrated in his *Miners' Friend* of 1702. A vessel equipped with a system of stopcocks was filled with steam, creating a vacuum as it cooled and thus sucking up water, which could be propelled further upwards by the force of the next blast of incoming steam. It was used in the water-works at York Buildings near Charing Cross in London to draw water from the Thames, but probably never in a mine. That big step was taken by Savery's fellow Devonian Thomas Newcomen (1664–1729), a blacksmith and Dissenter who, in 1712, made the first successful steam engine – but had to enter into partnership with Savery since a patent granted to the latter was deemed to include it, despite all Newcomen's innovations.

Newcomen's engine incorporated a great seesawing beam connected, on one side, to the cylinder through its piston and, on the other, to the pumps down the mine. The cylinder was filled with steam from the boiler, the supply was then cut off, and cold water injected to condense it. As the pressure inside the cylinder fell sharply, the atmosphere forced the piston down, rocking the beam and raising the pump rods. The hot water from the condensed steam was let out of the cylinder, fresh steam admitted as the beam rocked back again, and the cycle repeated as desired. The story was told that a boy, put to mind the engine, thought of connecting the stopcocks so that the machine became self-acting; there is no firm evidence for that, but certainly by 1717 Newcomen's engines no longer needed manual control of the taps. The engine had to be large so that the atmospheric pressure on the piston was sufficient to overcome all the friction and do the necessary work; but it was not possible to make a big iron cylinder with a bore that was precise enough for the piston to fit tightly within it. The piston was therefore sealed on the top with leather covered in water, which did not evaporate too speedily because the cylinder was heated up and cooled down at each stroke. If the engine were set to run on its own it might suck the sump at the bottom of the mine dry, when the powerful downstroke of the piston would fling the pumping machinery into the air. To guard against this, a timing device called a cataract, found in Islamic water clocks, was used: a controlled flow of water dripped into a cup on one arm of a little seesaw and after a certain time tipped it down, raising the other arm as the cup emptied itself. The raised arm would each time turn one of the taps and activate the engine, which would therefore pump at the preset rate.

Newcomen's engines were extremely inefficient chiefly because the large cylinder had to be heated up by the steam and cooled down by the water jet on each stroke. With experience, engineers – Smeaton being the most eminent – improved them by adjusting the dimensions and the timings, but the engines remained hungry for fuel. At the pithead, coal (especially dusty slack) was cheap, and without the engine no coal could have been won, and so expense was a matter of indifference. In Cornwall, where coal had to be brought in from Wales to drain the tin mines, it was different – and it was there that Watt's much more efficient engines came into use: they had separate condensers (always cool) and pistons fitting much more closely into the cylinder (always hot). His business partner, Matthew Boulton, in his Birmingham foundry developed the metal-working skills that made the good fit possible. Newcomen had been in contact with Hooke, and it is thus tempting to see the steam engine as an outcome of experiments on air pressure, but this does not seem to have been the case. The Huguenot refugee John Theophilus Desaguliers (1683–1744), a clergyman, protégé of Newton, promoter of Newtonian philosophy in public lectures and Copley Medallist, was snooty about Newcomen's ignorance, but described and illustrated his engine in his *Course of Experimental Philosophy* (1734), thus making it familiar in learned circles. But it was to be many years before men of science would, following Carnot, come to understand the conversion of heat into work, and there-after make further substantial improvements to steam engines.

Meanwhile, warfare and improving communications also required engineers, both military and civil; they were important participants in the Scientific Revolution. In 1453 the Turks, who had already built a fortress just north of Constantinople to control the Bosphorus, breached with their artil-lery the great walls that had for a thousand years protected the city. Against heavy gunfire, old-fashioned castles and high, thick stone city walls built to withstand sieges crumbled. In Italy, a new design for fortresses evolved: lower, squat, protected by a ditch rather than the old moats, and star-shaped rather than round or square. They were devised so that attackers would be raked by crossfire from guns in the star points, and by musket fire through slits. Engineers in England followed this pattern in fortifying the frontier town of Berwick-upon-Tweed against the Scots and their allies – this was one of the most expensive works of Queen Elizabeth I's reign (1558–1603). Their impressive walls still survive, never having been tested to destruction, unlike contemporary forts in the Low Countries. There, the main feature of the interminable wars between the Dutch, Spanish and French armies was the building, holding, taking and retaking of fortresses and fortified towns like Maastricht (where formidable works can still be seen).

The most distinguished exponent of the new science of siege warfare was Sébastien le Prestre de Vauban (1633–1707), who as a young man in a rebel army was captured and turned round by the forces of Louis XIV, became responsible for fortifying well over a hundred places in France and beyond its current borders, and in 1703 was created a marshal of France. His aim was to create a defensible border for northern France, including much of the previously Spanish Netherlands; but he was as famous for taking towns, including Ghent, Ypres and Maastricht, as for fortifying them. He invented a way of fixing bayonets still used in our light bulbs; devised a system of parallel trenches for attackers; and at the siege of Namur (1692) used ricochet fire to knock down walls. The walls he built at Strasbourg in 1681 remain an impressive feature of the city; and others of his fortresses, their menace diminished by weathering and the passage of time, are now UNESCO World Heritage sites.

Communications were as important as fortresses for military purposes as well as for general prosperity, and in 1716 the French government founded a Corps des Ponts et Chaussées responsible for roads and bridges. In 1747 a training school, perhaps the world's first civil-engineering college, was founded under Jean-Rodolphe Perronet (1708–94), who had designed a number of important bridges and who became responsible for the building and maintenance of more than a thousand miles (1,600 km) of roads. Meanwhile, in Britain the state of the roads had been interfering with commerce and travel, as they became increasingly muddy and rutted under the pressure of traffic. Each parish was supposedly responsible for the upkeep of its own roads, but from 1663 magistrates were allowed to put gates across main roads and charge a toll that would be used to maintain the highway. This was done for some lengths of the Great North Road that linked London to Newcastle and Edinburgh. From 1706, this power could be granted by Act of Parliament to private companies, called Turnpike Trusts, which might raise capital on the expectation of revenues from tolls. Under this privatised system nearly one-sixth of the roads were turnpiked over the course of the next 150 years. Engineers including Thomas Telford (1757–1834) devised better methods of construction, and built bridges. Though expensive, turnpikes greatly speeded up the mail, and reduced journey times for stage coaches and wagons; they were popular with travellers, but disliked by the poor, who were now expected to pay for using what had been free.

Even on the smoothest road, moving a heavy wagon took a great deal of energy. In the mining districts the old idea of moving trucks on rails was revived in the seventeenth century for getting coal from the pits scattered about the countryside to the staithes, where it could be loaded onto boats (keels) or ships (colliers) for transport to London and other cities. The rails were made of wood, and chauldron wagons with flanged wooden wheels were built to run on them, a 'chauldron' being a variable unit: 53 hundredweight (2.6 tonnes) in Newcastle, but in London about half that. On level ground or downhill a horse could pull about ten tonnes that way, and bring the empty wagons back up again; and a network of wagonways was built connecting mines to rivers, landowners charging a wayleave on wagons crossing their land. They had to be level to be practicable, and that entailed cuttings, embankments and bridges – work for engineers, carpenters and masons, as well as labourers. The oldest railway bridge in the world, Causey Arch in County Durham, has a span of over 100 feet (30 m) and was built in 1725–6 across the burn, or beck, more than 66 feet (20 m) below: for thirty years it was the longest single-span arch in Britain. The promoters were the Grand Alliance of coal-owners, and the builder was Ralph Wood, a mason, who used stone after his first wooden bridge had collapsed; despite his eventual triumph – the bridge still stands as an industrial monument and tourist attraction – in a fit of depression Wood threw himself from it to his death in 1727. At its peak the bridge carried some 930 chauldron wagons a day; but the pit in Tanfield blew up spectacularly in 1740, and by 1770 the bridge was becoming disused. Wagonways were subsequently to evolve into the first modern railways, as iron replaced wood, and steam engines (mobile or stationary) horses.

Wagonways carried their heavy loads to water, because boats and ships remained the most efficient means of transport. Rivers were important inland waterways, on which most great cities were built, and provided access to the sea. They are not ideal: flowing with variable speed through gorges and shallows, they meander through level ground, rush over alarming rapids (where boats may need to be carried, in portages), develop snags as banks cave in and trees fall, and are liable to flood or dry up. Improving upon them, the Chinese dug canals from very early times. Where canals could follow a contour, there were no problems, but going up- and downhill required the 'pound locks' with a gate at each end with which we are familiar, and these go back to tenth-century China. Water supply was also needed at the summits of such canals to keep them topped up, and this might require great works. In France, there had long been interest in the possibility of a canal that would form a short cut between the Mediterranean and the Atlantic, and King François I (1494–1547) sought Leonardo da Vinci's advice about it when, in 1516, he became his patron and brought him to France. In the following century, when tax collecting in France was 'farmed out' (privatised), the wealthy tax-farmer Pierre-Paul Riquet (1609–80) conceived, in 1662, a canal from the Garonne at Toulouse to the Aude at Carcassonne; and through Colbert, the chief minister, induced Louis XIV to back it because of its strategic and commercial importance. The king appointed the eminent soldier and mili-tary engineer Louis-Nicolas de Cherville (1610–77) to direct the project, which was on a grand scale: the Canal du Midi, as it was called, had forty aqueducts and ninety-one locks, and required complex dams and other water-works as well. The workforce over the many years of construction included large numbers of women. The canal was opened in 1681, after the death of both its prime backer and its chief engineer. It is now, like other great works of engineering, a UNESCO World Heritage site.

When Francis Egerton, 3rd duke of Bridgewater (1736–1803), whose estate at Worsley, north-west of Manchester, was rich in coal, visited France, he saw the Canal du Midi and was much impressed. He formed the idea of building a canal into Manchester, where there was a booming market for fuel, sought a suitable engineer, and through his steward engaged as consultant the millwright James Brindley (1716–72). Brindley had little formal education but was literate and had shown huge ingenuity in water-works. An Act of Parliament authorising the canal was duly passed in 1759; opened in 1761, it inaugurated a network embodying Brindley's vision, which soon linked the industrial and Midlands of England to London and other markets at home and abroad, and even crossing the Pennine hills to Leeds. Suitable for narrowboats, his canals were not as broad as those in continental Europe, where land was cheaper. Masons had long been involved in adventurous structures; Leonardo was always fascinated by water; military engineers had to think about moats and ditches; and mill-wrights installed water wheels. But in the days before engineering became a profession with different branches and formal training, the way was open for able young men like Brindley to take advantage of opportunities. Their work was certainly science in the form of Huxley's 'trained and organised common sense'; these projects depended upon experienced engineers, were big, bold and expensive, and were as crucial a component of the Scientific Revolution as the natural philosophy of Descartes and Newton.

Such fruits of science could be seen and admired, as could air pumps and microscopes; but science as public knowledge needed clearly set-out mathematics, diagrams, descriptions, and analyses of observations and experiments, and these had to be published. Often in the first instance that meant demonstrations to colleagues, students or other publics; or letters to friends, acquaintances and publicists like Mersenne and Hartlib. But then print publication was essential for further dissemination so that more people could join in. The Scientific Revolution was carried forward by the printers in the German lands, in Venice and then, especially, in the tolerant and commercial Netherlands; and their wares were distributed internationally through the annual Frankfurt Book Fairs. Scholars also met there, and in printing houses such as Plantin's in Antwerp. In more centralised France and Britain printing was regarded with suspicion by the powers that be, restricted, licensed and censored; and that applied also to the Spanish Netherlands after the reconquest (1585) when Antwerp lost its pre-eminence. Books and pamphlets printed in the Netherlands, sometimes with false publishers' names, were smuggled out into less tolerant countries: both licit and illicit publications involved not only text but also engravings – allegorical title-pages, illustrations of apparatus and maps. Lack of copyright across frontiers also favoured 'pirated' editions which competed with authorised publications. With its loose federal government and its Reformed religion, the Dutch Republic was a magnet for English Protestants fleeing Henry VIII and Mary I in the sixteenth century, and the high-church innovations of Charles I and Archbishop Laud in the seventeenth.

James Moxon was one of these refugees, going to Delft in 1636 and establishing himself in 1638 as a printer in Rotterdam, where he published Bibles and tracts for the English market. His son Joseph (1627–91) accompanied him, and learned the language and the trade before returning to London about 1647, bringing Dutch types with him. He was to base his career on what he had learned in the Netherlands, at a time when the

Dutch, the French and the British were competing to wrest command of the seas from the Spanish and Portuguese, and seize (or trade with) their colonies, and when censorship in Britain was in retreat.

In 1649 Joseph Moxon took up the study of geography and hydrography, with the necessary mathematics, and in 1652 went to Amsterdam, negotiating with an engraver about borrowing or preparing maps or illustrations for publication. By 1653 globe and map printing and selling had become the major part of his business, and in 1654 he translated Willem Blaeu's *Tutor* on the use of celestial and terrestrial globes. In 1659 he published his own book on globes, and in 1674 dedicated its third edition to Pepys. In 1662, despite his puritan background, he was appointed hydrographer to Charles II and got to know not only Pepys but also Boyle, Hooke and, later, Halley. After the Fire of London in 1666, when his shop was destroyed, he set up in Westminster, by now doing his own letter-cutting and type-founding rather than relying on Dutch type. He cut punches for the symbols in Wilkins' artificial language, and for Irish type wanted by Boyle; and in 1678 was elected to the Royal Society, in which one of the few other tradesmen in the seventeenth century was John Graunt (and he was nominated by the king). We get an idea of Moxon's mathematical and religious enthusiasms from his doggerel verses about Napier's Bones:

> Religious Romanists strongly maintain
> That by the Bones of their dead Saints are wrought
> Wonders; 'tis strange! Yet they the purses drain
> Of them that to their fond Belief are brought.

> But we'l reject those fancies, let them go
> With their dead Trump'ry, here's Lord *Napiers* Bones,
> Which Ile ensure you will more wonders show
> Then all those Reliques they count *holy Ones*.

> Canst thou but Add, then thou maist Multiply,
> And if Subtract, 'twill teach thee to Divide,
> And likewise to Gauge Vessels suddenly,
> And measure both Glass, Board, and Land beside.[6]

In January 1677/8 he published the first part of his monthly *Mechanick Exercises, or the Doctrine of Handy-Works*, a very early example of serial publication – something that became very common, for example, for Victorian novels. This was a full description of trades and tools, intended to demystify craft mysteries in a time when consumer demand was putting pressure on guild practices, and bodies such as the Royal Society and the Académie des Sciences were keen to learn about trades, publish accounts of their practices and perhaps give them rational principles.

When complete, *Mechanick Exercises* was published in two volumes, the second devoted to printing; there were several editions, the last of which came out in 1703, edited by Moxon's son James, who added a section on bricklaying, an important craft after the Great Fire of 1666, when it was made illegal to build in wood in London. The book belongs to a small but valuable genre of scientific publication that simply tells readers how to do things, without any theory: Michael Faraday's *Chemical Manipulation* (1827) and 'do-it-yourself' manuals are among its successors. Its sections concerned smithing, joinery, house-carpentry, turning, bricklaying and making sundials, each with its own recommended further reading, and glossary-cum-index. Within each section, the associated tools and activities were described. The book began with a preface justifying the enterprise:

> I See no more Reason, why the Sordidness of some Workmen, should be the cause of contempt upon Manual Operations, than that the excellent Invention of a Mill should be despis'd, because a blind Horse draws in it. And tho' Mechanicks be, by some, accounted Ignoble and Scandalous: yet it is very well known, that many Gentlemen in this Nation, of good Rank and high Quality, are conversant in Handyworks: and other Nations exceed us in numbers of such. How pleasant and healthey [*sic*] this their Diversion is, their Minds and Bodies find; and how Harmless and Honest, all sober men may judge.

He added that sciences including geometry, astronomy, drawing in perspective, music, navigation and architecture depended upon craft skills; and appealed to the authority of Bacon for the idea that much 'experimental philosophy' is involved in crafts, which might be improved 'by a Philosopher'. He also noted that practitioners of one trade might 'borrow many Eminent Helps' from another.[7]

Everybody knows that anything, and perhaps especially anything manual, is better learned from a person than from a book. But Moxon's, with its full descriptions and its illustrations, would be a most valuable accessory, and is a wonderful guide to how things were done. Thus the adze 'hath its edge athwart the Handle', like a hoe, making it more difficult to sharpen than an axe, and is used for smoothing:

> When they work upon the framed Work of a Floor, they take the end of the Handle in both their Hands, placing themselves directly before the Irregularity, at a small distance, straddling a little with both their Legs, to prevent Danger from the edge of the Adz, and so by degrees hew off the Irregularity. But if they hew upon an Upright, they stand directly before it. They sometimes use the Adz upon small thin Stuff, to make it thinner ... and then they lay their Stuff upon the Floor, and hold down one end of it down with the Ball of the Foot, if the stuff be long enough; if not, with the ends of their toes, and so hew it lightly away to their size, form, or both.[8]

This really is the language of artisans at which the Fellows of the Royal Society like Hooke aimed, though in his case wit kept bursting through.

At the back of Moxon's book there are advertisements for the globes, maps and books that he (or by now his son) had for sale in 1703. Moxon had financial ups and downs, failing to pay his Royal Society subscription in 1683 and leaving only £39 at his death. But the shop must have been competently run and stocked, responding to demand; and, among the globes of various sizes for sale, two (14 or 8 inches [35 or 20 cm] in diameter) displayed the Ptolemaic system, so there must still have been a market for them among conservatives. Moxon's second volume, on printing, depicts men (and sometimes disembodied hands) at work in thirty-three illustrations; has an appendix on the already ancient customs used in a printing house, called a chapel; and describes the processes of letter-cutting, typecasting and printing with old-fashioned and new-fashioned presses, and a screw-press for copperplates. He demonstrates how the practices vividly evoked in the Plantin-Moretus Museum in Antwerp were brought from the Netherlands to England during the seventeenth century. Moxon saw himself as a typographer, a job requiring the qualities of an architect:

> By a Typographer, I do not mean a Printer, as he is vulgarly accounted ... I mean such a one, who by his own Judgement, from solid reasoning with himself, can either perform, or direct others to perform from the beginning to the end, all Handy-works and Physical Operations relating to Typographie. Such a Scientifick man was doubtless he who was the first Inventer of Typographie.[9]

Moxon in his own eyes was no mere craftsman, but a man of science who compared himself to John Dee. Herbert Davis and Harry Carter's edition of *Mechanick Exercises* lists all the books, mostly on geography and mathematics, that Moxon published; but it is for the *Mechanick Exercises* itself, realising a dream of the Royal Society's founders, that he is remembered. In 1728 Ephraim Chambers (1680–1740) published its successor, his *Cyclopedia*, a two-volume compendium of technical information, and this was the spur that set Denis Diderot (1713–84) and Jean le Rond d'Alembert (1717–83) to work, compiling the great *Encyclopédie* which is seen as the Enlightenment embodied. As well as sly inflammatory pieces on church and state, the latter work has wonderful plates and descriptions of arts and crafts in France in the mid-eighteenth century, and represented the culmination of a project to study trades begun by the Académie des Sciences in its early days. By then the adze was obsolete, but we see many small figures in great activity making both useful everyday things and articles of luxury and elegance.

Crafts, trade, machinery and printing made the modern world, but farmers fed it; and until the nineteenth century the great majority of people in Europe worked on the land. Most of our ancestors were peasants; but as land previously cultivated communally in strips was enclosed and transformed into private property, tenant farms, rack rents, cash crops and day-labouring became more normal from the later seventeenth century onwards. An increasing urban population wanted more varied food, and more of it.

Parisians by the eighteenth century were fastidiously demanding white bread in place of coarser (and more nutritious) wholemeal, and were prepared to riot for it. The prosperity of the Netherlands in the seventeenth century, and of France and Britain in the eighteenth, resulting from foreign trade, was accompanied by an associated agricultural revolution. Speedily tobacco and gradually potatoes and corn (maize) found their way into European habits and diets, and Dutch prosperity meant demand for meat, butter and cheese. The latter imported wheat from the Baltic, along with the timber, hemp and tar needed for their ships, and kept cows. They rapidly improved the breeds (as the ubiquitous Friesian indicates), and manure from the cowsheds fertilised the soil in land being steadily reclaimed from the sea, so that not only tulips but also large crops of vegetables could be grown in what came to be called 'high farming'.

This practice spread to England with many other Dutch ideas, fashions and financial institutions after the 'Glorious Revolution' of 1688 brought William and Mary to the throne. The process of enclosure accelerated. Great landowners began to dispossess peasants and charge the high rents that enabled them to build and landscape great houses but required tenant farmers to become more efficient. About 1700 Jethro Tull (1674–1741) invented both a seed drill that planted seeds in rows, and a horse-drawn hoe that would weed between them, and described them in his *Horse-Houghing Husbandry* (1731). Mineral fertilisers like marl (clay mixed with lime, calcium carbonate) supplemented farmyard manure. Crops increased, but farm labouring remained back-breakingly hard work and harvests were always uncertain. Potatoes came in gradually to supplement bread all over Europe, and in Ireland in the course of the eighteenth century steadily became the staple food of the poor.

The other great feature of this agricultural revolution was stockbreeding. The prominent statesman Charles Townshend (1674–1738) was celebrated in his retirement for introducing turnips on his estates in Norfolk as a winter feed for cattle. Enclosure, and the introduction of turnips, meant that more controlled mating was possible, and autumnal slaughter no longer inevitable. Different breeds of sheep and cattle suited more or less to different conditions around Europe had emerged, but breeding had been random rather than selective: here was scope for the experimental spirit to move from the laboratory to the field. Hounds, sheepdogs, turnspit dogs and lapdogs had long been carefully bred; heavy horses too had been bred for farming and for knights in armour, and more elegant mounts for ladies and gentlemen to ride. The Habsburgs had promoted the breeding and training of horses for dressage, and in central Europe Arab horses were growing famous for their speed and endurance as horse-racing became a craze. In 1616 King James I imported the 'Markham Arabian' to Britain. In 1686 the 'Byerley Turk' was captured (from the Turks) at Buda, and early in the eighteenth century the 'Darley' and 'Godolphin' Arabians (named for their owners) were imported into England: from these, the bloodlines of racehorses are still traced with the loving care and attention devoted to aristocratic human pedigrees. They had their portraits painted too: George Stubbs (1724–1806) based his on careful anatomical studies, dissecting horses in an isolated farmhouse and indifferent to the smell, for his magnificent folio *Anatomy of the Horse* (1766). Perhaps partly to pay the bills for maintaining racing stables and debts incurred in betting, landowners and their tenant farmers began to show similar care in breeding their cattle and sheep, and yields of milk, beef, pork, wool, lamb and mutton improved as best practices were spread through county agricultural shows. The Spanish merino sheep was famous for its fine fleece, and their export was forbidden to protect the wool trade. Some were nevertheless smuggled out, but they are small, and in Britain not only is the climate different, but sheep farming depended for profitability on the sale of lamb and mutton as well as wool. The stockbreeder had to balance the various needs of markets in choosing which animals to use for mating.

Farming, like everything else, was being changed in consequence of, and as part of, the processes that gave us the Scientific Revolution. And the interest in domestic animals and edible vegetables was accompanied by an enthusiasm for natural history, especially as voyagers brought back exotic plants and animals that might be useful, and were certainly to be wondered at. It is to that field, again just as important as astronomy and much more accessible, that we now turn.

CHAPTER 10

THE LADDER OF CREATION
THE RISE OF NATURAL HISTORY

THE WORLD IS so full of wonderful things, animal, vegetable and mineral, that humans everywhere have always delighted in it and sought to make sense of it as well as to master it. European voyagers from the 1490s on crossed the 'torrid zone' to temperate southern Africa, or the Atlantic to Mexico, Peru, Brazil, Florida or Virginia. They were amazed to find there in climates very similar to those of Europe animals and plants quite distinct from those they knew at home – and yet alike in plan despite all the differences in detail. The God of the *Benedicite*, the biblical 'Song of the Three Holy Children' sung in Nebuchadnezzar's burning fiery furnace, clearly rejoiced in variety, and natural history went readily with natural theology. All peoples had devised classifications: useful (game/vermin, tame/wild); defining a group (kosher, halal, vegan) and in casuistic deference to religious precepts (if geese grew from barnacles, monks might eat them on Fridays as fish); or theoretical, a taxonomy based on differences and resemblances perceived or inferred, more or less rigorously bifurcating or based on family groupings. Pliny's *Natural History* (c. 77) was a great resource for medieval scholars and on through the Renaissance; Aristotle's zoological works with their emphasis upon reproduction and function became extremely influential; and Dioscorides' *Herbal* (first century AD) was a source for later botany.

It gradually became clear that the plants and animals of northern Europe were not the same as those described by these authorities from the Mediterranean; notably to William Turner (c.1510–68) from Northumberland in the far north of England who studied at Cambridge, became a red-hot Protestant and travelled widely in order to escape King Henry VIII's clampdown on heretics. In Italy, at Ferrara and Bologna, he studied medicine and botany; going north, he became a friend of Conrad Gessner (1516–65), the Zurich polymath and encyclopaedist, before practising as a physician in Friesland. Home again, pursuing a career as both physician and clergyman, he advised the sympathetic new archbishop, Thomas Cranmer (1489–1556), about contacts with foreign Protestants. Turner published the first original English book on plants, *Names of Herbes*, in 1548, and *A New Herball* in 1551–62, the latter's publication being interrupted by further travels in Germany, to Cologne and Worms, during Queen Mary's brief restoration of Catholicism. A patriot, he wrote: 'I might in my herbal declare to the greate honoure of our countre what numbre of souereine & strang herbes were in Englande that were not in other nations.'[1] He was an all-round naturalist, providing information on fishes for Gessner, and publishing close observations of birds and their behaviour. His *Herball* used woodcut illustrations from Leonhard Fuchs (1501–66), of the Protestant university at Tübingen, who was depicted in 1542 describing a plant in a vase while his draughtsman drew it: nature rather than authority was to be the guide for scholars. Turner's work shows the importance of travel and making close connections across frontiers, as well as the complex way that the Scientific Revolution and the Reformation became intertwined.

Botany was studied and published as an offshoot of medicine, the profession of Fuchs, Turner and Gessner, though the influx of new plants from America, Africa and Asia also led to a boom in horticulture, bringing new fashions in garden design, with magnolias to delight the eye as well as potatoes to fill the stomach. Gessner, who was educated in Strasbourg, Montpellier and Basel, and loved his native Swiss countryside and Alpine plants, was most eminent in his own time as a botanist, but is now chiefly remembered for his *Historia Animalium* (1551–8), with its wonderful woodcut illustrations of both familiar and exotic animals. His rhinoceros came from Dürer and so looks even more splendid than a real one, his elephant has magnificently pleated ears, and his lion has a kingly and almost benign air: these are appropriate bits of artistic licence, for animals had symbolic importance going back to Aesop's fables (sixth century BC) and similar stories in other cultures, and fashionable in Renaissance emblem books. But Gessner's dragons and unicorns, also highly symbolic heraldic creatures, do not exist except in art and literature; and, like Pliny, Gessner took his authorities more seriously than did his successors, who were armed with the Royal Society's motto, 'Nullius in verba'. Because nature is so diverse, and so much of the world was unknown to Europeans even in 1776, it was impossible to be certain that some creature or behaviour was chimerical – as indeed it still is.

Gessner was the major source for the ambitious London clergyman Edward Topsell (1572–1625) who, in the tradition of medieval bestiaries, recommended his prolix *Historie of Foure-Footed Beastes* (1607; *Serpents* followed a year later) for Sunday reading:

> This is my indeavour and pains in this Book, that I might profit and delight the Reader, whereinto he may look on the Holiest daies. . . and passe away the Sabbaths in heavenly meditations upon earthly creatures. I have followed D[r]. *Gesner* as neer as I could, I do profess him my Author in most of my Stories, yet I have gathered up that which he let fall, and added many Pictures and Stories. . . . He was a Protestant Physician (a rare thing to find any Religion in a Physitian) although St. *Luke* a Physician were a writer of the Gospell. His praises therefore shall remain, and all living Creatures shall witnesse for him at the last day. This my labor whatsoever it be, I consecrate to the benefit of our *English* Nation.[2]

When, fifty years later, a new edition of the book was published, it came with another volume appended – *The Theater of Insects* by Thomas Mouffet (1553–1604), whose daughter Little Miss Muffet had a famous encounter with a spider. Mouffet was full of admiration for house spiders, from whom we could learn lessons of diligence and whose beauty we (unlike his daughter) could enjoy.

As well as spiders, Mouffet's category of 'insects', or lesser living creatures, included scorpions and millipedes; but other authors used the term in a much wider sense. Thus, in his *New Voyage to Carolina* (1709), the surveyor John Lawson (d. 1711) listed the insects found there. This category consists entirely of what we would call reptiles, including alligators and snakes. Lawson justified himself as follows: 'Tortois, vulgarly call'd Turtle; I have ranked these among the Insects, because they lay Eggs, and I did not know well where to put them.'[3] Elsewhere, he put tortoises among the 'Shell-Fish'. Lawson's categories may have been eccentric, but they were also lasting: well over a century later, the Victorian naturalist Frank Buckland (1826–80), embarking on a railway journey, was told that he must pay for his monkey because it was a dog, but not for his tortoise because it was an insect. Classifying, placing, is not straightforward, whether its purpose is scientific or commercial.

Topsell never dealt with birds, though Gessner had done so; and so had Pierre Belon (1517–64), who had travelled widely in the Middle East, and whose book on the nature of birds (1555) included a plate illustrating the homology of the avian and human skeleton. The *Ornithologia* of his contemporary Ulisse Aldrovandi (1522–1605) was handsomely illustrated and includes a charming woodcut of young nightingales in a nest, which in some copies is unfortunately printed upside down. Trained in medicine and law at Bologna and Padua, in 1549–50 he was charged with heresy and put under house arrest in Rome; there, he became enthused with natural history, and on being freed returned to Bologna, where he was made a professor, founded the botanic garden in 1568, assembled a large herbarium of dried plants and built up a splendid cabinet of curiosities with thousands of specimens. Like Gessner, he was an encyclopaedist rather than a critical naturalist; but by consolidating knowledge from the past and the present, they laid the foundations upon which others would build.

One of the sceptics sieving such works was the physician Sir Thomas Browne (1605–92), who had studied at Oxford, Montpellier, Padua and Leiden, and practised in Norwich. In 1646 he published, in his characteristic witty and magnificent Latinate English, his longest book, *Pseudodoxia Epidemica*, being 'enquiries into very many received tenets and commonly presumed truths'. He found mankind credulous and supine, readier to follow rhetoric than logic, too easily accepting of authority, open to temptation by Satan and too prepared to believe hucksters, 'Saltimbancoes, Quacksalvers, and Charlatans'. Among the things Browne found incredible were tales of the phoenix, or the amphisbaena (a snake with a head at each end), or of gryphons, basilisks, cockatrices and dragons. He rejected also the widely accepted notions that badgers have legs that are shorter on one side than on the other, that hares change sex, that elephants have no joints and must sleep leaning on a tree, and that 'a Bever, to escape the Hunter, bites off his testicles or stones'.[4] These, for him, were fables akin to proverbs, and made no sense when applied to actual animals. His criticisms were not based on his own observations, but rather on logic and on modern accounts by authors he trusted. Again, he was not altogether successful in abolishing inaccuracy: there is a wonderful illustration of a hairy two-headed serpent (assailing a monkey with one head and a deer with the other) in the superb series of paintings of life in Peru commissioned by Martinez Compañon (1737–97) after he was appointed a bishop there in the 1780s.[5] The fearsome creature, which must be at least 20 feet (6 m) long, is contemplated by an astonishingly phlegmatic peasant standing perilously near it. This image, it must be said, jumps out at the modern reader from its place in an otherwise factual, if frequently stylised, series of paintings by Peruvian artists of plants, animals, people and towns.

Aristocrats and scholars in the Renaissance delighted in collecting, building up cabinets of curiosities to inspire wonder, to show off and confirm status, or to make a talking point for guests. Exotic objects, natural or artificial, from new-found lands across the seas were prominent, as sailors and merchants brought home curios to sell, or on commission. Rudolph II's collection, displayed in his *Wunderkammer*, or chamber of marvels, became particularly famous. Only the privileged might be allowed access to those of grandees, displaying amazing natural objects and freaks, and enriched with antiquities, gems and valuable artefacts admired for their beauty and ingenuity. Later generations found it hard to make sense of such cabinets: they seemed higgledy-piggledy, the fruit of a childish urge to collect miscellaneous things. But modern historians have discovered them to be a neglected but fruitful source of knowledge, for, where they, their catalogues or reports of them are preserved, they give us an insight into the habits and interests of a lost encyclopaedic age of wonder. They were open-ended in the way that the best museums today can be, provoking astonishment and delight.

The more specialised collections beginning to be formed by scholars had a serious purpose, the close study and comparison of specimens; and natural history came to mean forming and carefully examining focused collections, corresponding with other enthusiasts, swapping duplicates, trying to fit them into a taxonomic pattern and publishing an illustrated work about them. Minerals, fossils and shells, being almost indestructible, were the easiest things to collect and preserve; moths and other insects were all too likely to attack stuffed animals and birds, but specimens preserved in jars of alcohol lasted well. Plants were pressed and stored when dry in a herbarium or *hortus siccus*, labelled and annotated; their colours drained away but their forms were preserved. As time went by, naturalists came to spend most of their time indoors classifying creatures they had never actually seen in the wild.

Such scholarly 'cabinet zoology' did not satisfy Martin Lister (1639–1712), who in 1678 published a book on English spiders. He described from delighted and careful observation their hatching and development, how they spun their webs, their feeding and their coition. Classical scholarship and deference to ancient authorities were no longer deeply relevant; and, again, while voyages to unknown lands gave a huge fillip to natural history, one effect was to stimulate closer study like this of European animals and plants. Lister's father, a Member of Parliament, had sent his son to Cambridge; in 1663 he then went on to Montpellier. Protestants were no longer able to matriculate there, but he could take part in informal discussions and hear lectures; after his return he married well and from 1670 practised medicine in York. In 1671 he was elected FRS, and he became a friend and correspondent of John Ray; in 1683 he moved to London and, chosen as vice-president of the Royal Society, took the chair when Pepys was absent (as, being prominent and busy, he often was). Gifts he made to the new Ashmolean Museum at Oxford University led to his being made a Doctor of Medicine there, which in turn allowed him to become a Fellow of the Royal College of Physicians. He participated in the postmortem on King Charles II, and in 1702 became one of Queen Anne's physicians. In 1697 his book on shells, *Historia Conchyliorum*, was published with over a thousand illustrations engraved on copper by his two eldest daughters; and it was for this authoritative work on the most indestructible and readily collectible specimens of natural history that he is chiefly remembered. Woodcuts had the advantage that they were relatively cheap to produce, and could be set with type so that picture and text could appear on the same page. But only copper engraving or etching could show the detail required to make the fine distinctions that Lister and others needed by this time. Although copperplates, an intaglio process, were expensive, needed a different kind of printing press and preferably rather different paper from type (a relief process), got worn on long print runs and had to be bound into books separately, they became the vehicle of choice for scientific illustrations until well into the nineteenth century. They were either printed black and white with careful crosshatching to indicate shading, or more lightly engraved to be hand-coloured as required following (with varying degrees of accuracy) a pattern plate prepared and often annotated by the artist. Where these survive, they are usually the nearest thing we have to an 'original' of the print.

Plants in natural histories today may be separated into trees, shrubs and herbs (as in Antiquity), edible and poisonous, or cultivated and wild; and flowers may, to assist in recognition, be arranged by colour, ecologically by climate or altitude, or decoratively by rarity and splendour. Much the same applies to animals, where a book may be systematic, covering a natural group like the owls, or ecological, featuring the mammals, reptiles, birds and insects of a region, either with carefully detailed illustrations or as a field guide with cartoonlike sketches to assist rapid identification. Plants and animals have local names, connected with folklore, and more widely used vernacular names. For Ray and his companions, it was hard to be sure whether different names, like nightjar and nighthawk, referred to one or more distinct kinds of creature; or conversely whether the same name might, like sycamore or robin, be used for different species. Scholars used Latin names, either seeking to identify the species named by ancient authors or coining fresh names; in a sentence or two they would describe the plant so that others could recognise it. But the scientific basis for this was uncertain, and Ray was a pioneer in applying an Aristotelian method of grouping in families rather than the strictly bifurcating 'Platonic' system which led to humans being described as 'featherless bipeds'. As we all know from weddings and funerals, members of families look alike in a general way, but no individual has the full list of family characteristics: nose, ears, hair, hands, deportment and so on. So it is with plants. Naturalists cannot simply follow rule of thumb, because appearances are deceptive: they need wide experience in order to spot important features and develop an intuitive feeling, akin to connoisseurship, of where plants belong. They must be able to disregard the individual peculiarities that we and all creatures display and to see their specimen as an example of a type. This generalising is important for the artist too when depicting a species rather than a particular member of it: *Rosa chinensis* rather than the rose from our favourite bush or *Psittacus badiceps* rather than Polly the parrot. Family grouping looks messy: some classes are much bigger than others, and there are puzzling or lonely specimens that could go into sundry different families, or may be 'nondescripts' with hitherto-unknown kin. The tidy-minded have throughout history sought patterns, but Ray, having found that Wilkins' idea of dividing up all nature into threes did not work, grouped by patiently taking as many characteristics as possible into account – in fact, one has to give more weight to some, like the form of flowers, than to others, like colour.

Such a method was not easy to teach. Like apprentices acquiring a mystery, or young gentlemen acquiring connoisseurship, pupils were to sit beside the expert, watching carefully until they had absorbed how it was done. Pithy, informed Latin descriptions of the various plants could then be penned and function as their scientific names. Ray's expertise across natural history was formidable, while zoology (especially of birds and fishes) was the forte of his patron, collaborator and friend Francis Willughby. Willughby's premature death in 1672 meant that Ray was left to complete and publish the resulting volumes on fish and birds. The handsomely illustrated *Ornithology* was published in Latin in 1676, with the financial support of Willughby's family. Two years later Ray published an expanded English edition

> to the illustration of God's glory, by exciting men to take notice of, and admire his infinite power and wisdom displaying themselves in the Creation of so many *Species* of Animals; and to the assistance and ease of those who addict themselves to this most pleasant, and no less

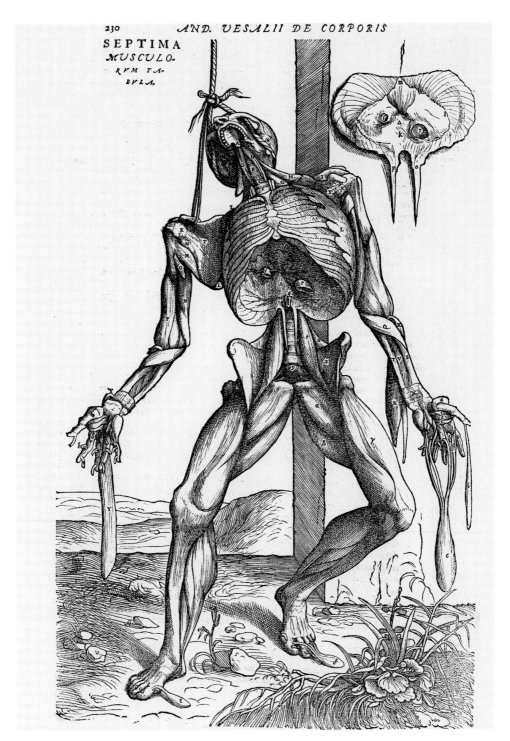

SEPTIMA
MUSCULO-
RVM TA-
BVLA.

30. Andreas Vesalius, *De humani corporis fabrica*, 2nd edition, that belonged to Bishop Cosin (Basel 1555). The pictures show the progress of dissection of a felon's corpse, displayed against a Tuscan landscape rather than in the anatomy theatre. In this plate, the cadaver has been stripped almost to a skeleton.

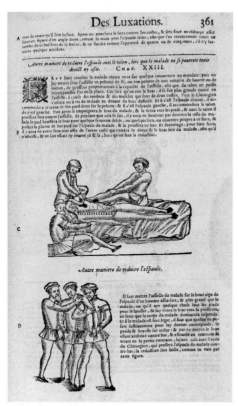

31. Ambroise Paré, *Les oeuures d'Ambroise Paré, conseiller et premier chirurgien du roy. Dixiesme edition, reueue et corrigée… & augmentée d'vn… traicté des fievres* (Lyons, 1641). A great military surgeon, Paré is remembered for abandoning both red-hot cautery after amputation and the use of boiling oil for gunshot wounds. Here he shows how to treat a dislocated shoulder.

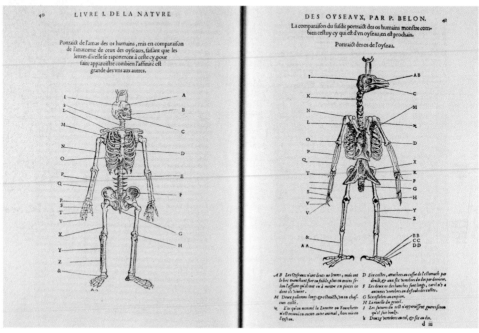

32. Pierre Belon, *Oiseaux* (Paris, 1555). The illustration, depicting the skeletons of a man and a bird, is a pioneering piece of comparative anatomy, showing homology, the pattern running through living creatures.

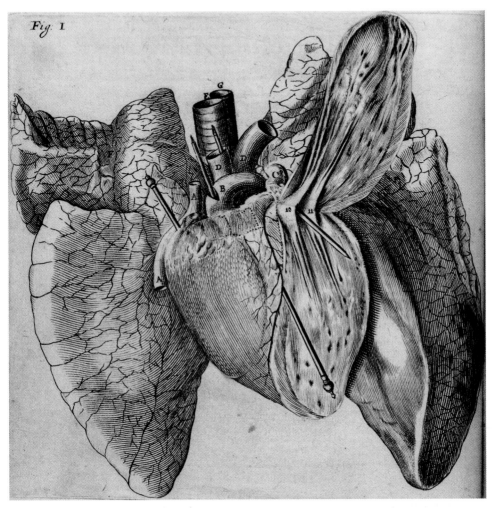

33. René Descartes, *De homine* (Leiden, 1664). Posthumous pop-up book with lifting flaps to show the interior of the human body. Descartes was a man of action and an experimentalist as well as a philosopher, in the Renaissance tradition.

34. Rembert Dodoens, *Cruyde boeck … Van nieuws ouersion, ende met seer veel schooner vieuwe figueren vermeedert* (Antwerp, 1563). The allegorical, classical title-page showing the Garden of the Hesperides, with various gods and heroes from Antiquity (rather roughly hand-coloured).

35. John Gerard, *The Herbal.* The 1633 edition includes the new plants maize (illustrated), sunflowers, tobacco and potatoes. The editor was a surgeon, and herbals were important in medicine.

36. Ulisse Aldrovandi, *Ornithologiae* (1637–46). A standard compilation on the birds of Europe and further afield. The illustration shows nightingale fledglings in their nest – upside down in this copy!

37. John Parkinson, *Paradisi in Sole Terrestris* (1656), frontispiece. The title of this herbal plays on the author's name, Park-in-sun; the plate shows the Garden of Eden, with the 'vegetable lamb of Tartary' (possibly a misunderstood cotton plant) grazing in the middle ground.

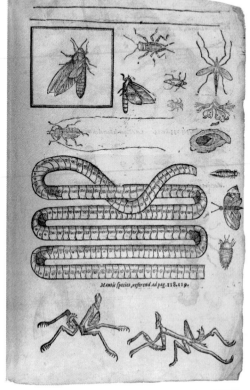

38. Thomas Moufet, *Insectorum sive minimorum animalium theatrum* (London, 1634). Mostly a compilation, drawn chiefly from Gessner. Moufet's daughter is said to have been the original of 'Little Miss Muffet'. A good example of pre-microscopic natural history, when 'insects' and 'reptiles' (our creepy-crawlies) were often run together into one great group; his praying mantises are very lively.

39. Athanasius Kircher, *Mundus subterraneus*… 2 vols (Amsterdam, 1665). Earthquakes and volcanoes were the subject of much speculation in the seventeenth century, and Kircher boldly investigated them. His cutaway illustration of Vesuvius is particularly splendid.

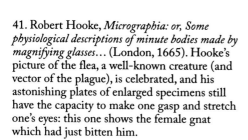

40. Francis Willughby and John Ray, *The ornithology of Francis Willughby ... Wherein all the birds hitherto known, being reduced into a method sutable [sic] to their natures, are accurately described ... To which are added, three considerable discourses ... by John Ray* (London, 1678). Ray and Willughby looked inside as well as outside their birds (the plate shows a dissection), and separated the sea-swallows (terns) from the swallows, despite their rather similar appearance. Some foreign birds which they had received as dried skins they depicted as such.

41. Robert Hooke, *Micrographia: or, Some physiological descriptions of minute bodies made by magnifying glasses...* (London, 1665). Hooke's picture of the flea, a well-known creature (and vector of the plague), is celebrated, and his astonishing plates of enlarged specimens still have the capacity to make one gasp and stretch one's eyes: this one shows the female gnat which had just bitten him.

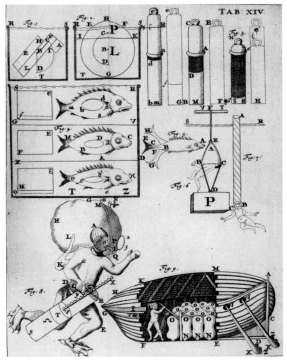

42. Giovanni Alfonso Borelli, *De motu animalium … Additae sunt J. Bernouilli meditations mathematicae de motu musculorum* (Leiden, 1710). First published in 1680, Borelli's *De motu animalium* presented the application of mechanics to the motion of the limbs of animals. Borelli thought that nerve stimulation was related to the contraction and swelling of a muscle and that some chemical process was associated with it. He also believed that the heartbeat was a simple muscular contraction and that the circulatory system was hydraulic in principle. This plate includes diving apparatus, a submersible boat, and how fish swim.

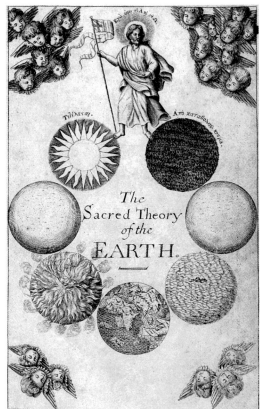

43. Thomas Burnet, *Sacred Theory of the Earth* (1684). The frontispiece shows the transformations of the Earth, past and to come, by water and fire; Noah's ark, alone amidst the waves, is at the bottom right. The risen Christ presides over this history.

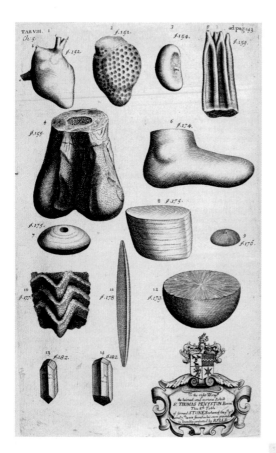

44. Robert Plot, *The Natural History of Oxford-shire* (1705). A plate of 'fossils' sponsored by a local landowner: note the flint in the form of a human foot. Fossil meant something dug up.

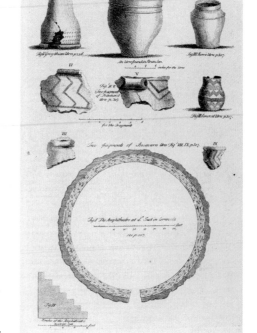

45. William Borlase, *Antiquities of Cornwall*, 2nd edition (1769). Plate illustrating archaeological site and finds, attributed to Druids. Borlase was interested in both natural history and antiquities.

46. John Ellis, *An Essay towards a Natural History of the Corallines* (1755). The corallines (illustrated here) were 'zoophytes', on the border between plants and animals and hence of ambiguous status. Ellis realised that they were definitely animals. One test was that while vegetable matter heated to destruction smells bad, animal matter smells horrible. His illustrations were done by the great botanical artist Georg Dionysius Ehret.

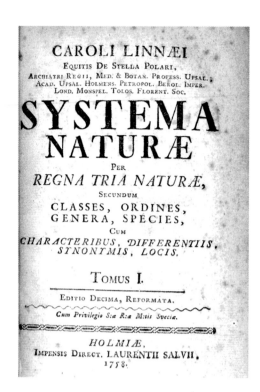

CAROLI LINNÆI
EQUITIS DE STELLA POLARI,
ARCHIATRI REGII, MED. & BOTAN. PROFESS. UPSAL.;
ACAD. UPSAL. HOLMENS. PETROPOL. BEROL. IMPER.
LOND. MONSPEL. TOLOS. FLORENT. SOC.

SYSTEMA NATURÆ

PER

REGNA TRIA NATURÆ,

SECUNDUM

CLASSES, ORDINES,
GENERA, SPECIES,

CUM

*CHARACTERIBUS, DIFFERENTIIS,
SYNONYMIS, LOCIS.*

TOMUS I.

EDITIO DECIMA, REFORMATA.

Cum Privilegio S:æ R:æ M:tis Sveciæ.

HOLMIÆ,
IMPENSIS DIRECT. LAURENTII SALVII,
1758.

47. Carl Linnaeus, *Systema naturae … Ed. decima, reformata*, 2 vols (Stockholm, 1758–59). The frontispiece: the scientific nomenclature of animals begins with this edition, and no earlier name – even if used earlier by Linnaeus himself – is now valid in science. Linnaeus' ideal of an international descriptive language was a crucial feature of the Enlightenment, and a great influence, for example, on Lavoisier and his associates as they tidied up chemistry.

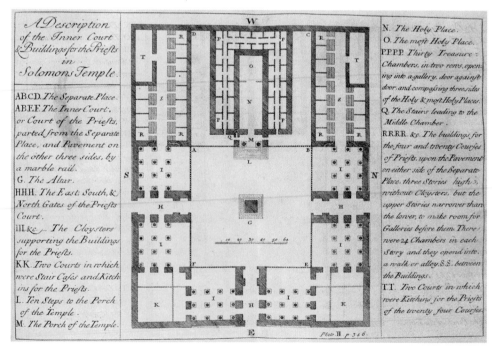

48. Sir Isaac Newton, *The chronology of ancient kingdoms amended. To which is prefix'd, A short chronicle from the first memory of things in Europe, to the conquest of Persia by Alexander the Great* (1728). Newton's reconstruction of the Temple of Solomon. He was very interested in Biblical interpretation, dating and prophecies.

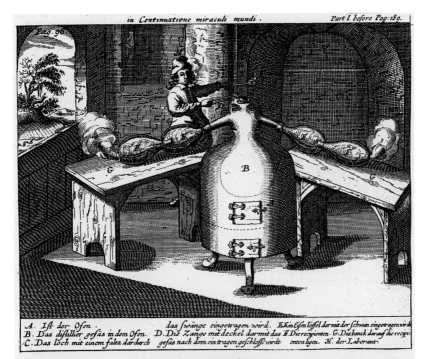

A. Ist der Ofen .   das swänge eingetragen wird .   E. kin Eisen löffel darmit der schwan eingetragen wird
B. Das distillier gefäs in dem Ofen . D. Die Zange mit deckel darmit das II. Die recipienten . G. Die banck darauf die recipi-
C. Das loch mit einem falz dardurch   gefäs nach dem ein tragen geschloffe wird: enten ligen . H. der Laborant.

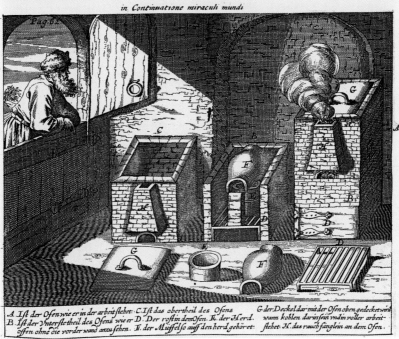

A. Ist der Ofen wie er in der arbeit stehet . C. Ist das obertheil des Ofens   G. der Deckel dar mit der Ofen oben gedecket wird
B. Ist der Unterste theil des Ofens wie er D. Der rost in dem Ofen . E. der Herd .   wann kohlen darinsein und in voller arbeit-
offen ohne die vorder wand anzusehen . E. der Müffel so auff den herd gehöret: stehet . H. das rauch fanglein an dem Ofen .

49. Johann Rudolph Glauber, *The works of the highly experienced and famous chymist ... containing, great variety of choice secrets in medicine and alchymy in the working of metallic mines, and the separation of metals ... Translated by Christopher Packe ...* (London, 1689). Germany had been devastated by the Thirty Years' War and Glauber and others hoped that revived mining and other proto-chemical industries could bring renewed prosperity. The picture shows distillation, and furnaces supervised from a safe distance.

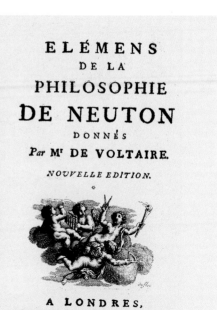

ELÉMENS
DE LA
PHILOSOPHIE
DE NEUTON
DONNÉS
Par Mr DE VOLTAIRE.

NOUVELLE EDITION.

A LONDRES,

M. DCC. XXXVIII.

ISAAC NEUTON
Gravé d'Après la Medaille &c.

50. Francois Marie Arouet de Voltaire, *Elémens de la philosophie de Neuton* [*sic*] *donnés par Mr. de Voltaire. Nouvelle edition* (Londres [i.e. Paris], 1738). With engraved frontispiece portrait of Newton. This book and the translation of Newton's *Principia* by Voltaire's mistress, Mme de Chatelet, made Newtonian science generally acceptable in France. Á Londres for place of publication is untrue; Voltaire was seeking to elude censorship.

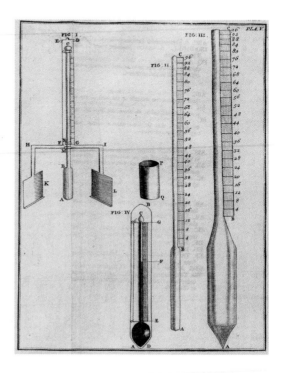

51. Herman Boerhaave, *Elements of chemistry: being the annual lectures of Herman Boerhaave ... / Translated from the original Latin, by Timothy Dallowe* (London, 1735). Boerhaave performed thousands of *in vitro* laboratory experiments and took down hundreds of case histories of patients in an effort to find new drugs and to understand the causes of disease. This is the authorised version of the famous chemistry lectures that he gave annually at Leiden: the plate illustrates experiments on heat.

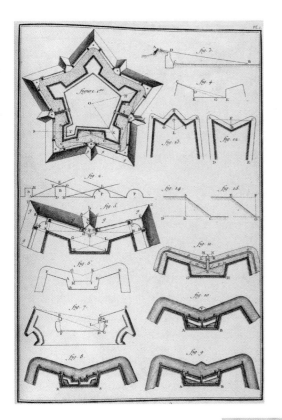

52. Denis Diderot, *Encyclopédie: ou, Dictionnaire raisonné des sciences des arts et des metiers … mis en ordre & publié par Diderot (and others)*, 35 vols (Paris, 1751–80). There are over 3,000 plates in Diderot and D'Alembert's great encyclopaedia: 'Art of War VI' shows modern fortifications, squat and built to mount and withstand artillery.

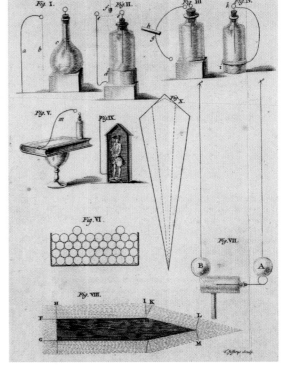

53. Benjamin Franklin, *New experiments and observations on electricity. Made at Philadelphia*, 3rd edition (London, 1760). Franklin's fame as a scientist rests chiefly on his electrical experiments. Flying a kite in a thunderstorm is extremely dangerous; an attempt to repeat it in Russia led to a vacancy in the Academy of Science. Electricity had meant parlour tricks: Franklin demonstrated its connection with very powerful forces.

54. Pierre Joseph Macquer, *Elements of the theory and practice of chymistry*, translated from the French of M. Macquer, 2 vols (London, 1775). Macquer's writings proved to be very popular not only for students, but with artisans, physicians, miners, farmers and other professionals whose livelihood depended upon chemical analysis. A student has used a blank page at the end for notes on a lecture by Joseph Black.

55. Joseph Priestley, *The history and present state of discoveries relating to vision, light and colours* (London, 1772). An early example of a chronological chart of contributors to the subject: Priestley got his Edinburgh doctorate for inventing the timeline. In his science, Priestley typically moved from history into research, from optics to electricity and then gases.

useful, part of philosophy; and also to the honour of our Nation, in making it appear that no part of real knowledge is wholly balked and neglected by us.

Ray added that they had wholly omitted '*Hieroglyphics, Emblems, Morals, Fables, Presages,* or ought else appertaining to *Divinity, Ethics, Grammar,* or any sort of Humane Learning: And present . . . only what properly relates to their Natural History'.[6]

While acknowledging debts to Gessner and Aldrovandi, Ray remarked that they had relied too much on ancient authorities and on careless informants; and in contrast to their bookishness he paid tribute to Leonard Baldner (1612–94), who described and illustrated the fauna of the Strasbourg region. Baldner was a fisherman, toll collector, head forester and finally councillor in Strasbourg, an excellent observer and dissector, whose manuscript Ray and Willughby had bought on their Grand Tour, and from which they had taken both information and pictures. In another version of the manuscript now in Cassel, the birds are depicted in colour, sometimes with foliage in the background: they look stiff because the specimens were dead but their poses are not unnatural, while the fishes, also coloured, are painted on top of the water as though floating. Ray also gratefully recognised his more learned contributors, Browne and Lister, Ralph Johnson (1629–95), a Yorkshire vicar and keen botanist, and two landed gentlemen, Francis Jessop (1638–91) and Philip Skippon (1641–91), who had all provided observations and pictures. In the book, after a general discussion, birds are divided into land and water types and grouped in families; we learn the weight and measurements of their specimen, what was in its stomach, and its colouring. Some of the engravings, which are uncoloured with several species shown to a page, are lively: the hen harrier clutches a bird that it has caught, the hobby stands atop a frog and the parrot cracks a nut. Exotic birds include the ostrich, the penguin, the toucan and the dodo, but the great majority are European. The basis for the classification is explained; this is unambiguously a work of science.

Historia Piscium, in Latin also, was published in Oxford in 1686, at the expense of the Royal Society (which then had no money left to subsidise Newton's *Principia*), and has a magnificent frontispiece dedicated to Pepys as president, and fine engraved plates. Fishes are not an attractive subject for the artist: on the slab they look very dead, different kinds appear rather alike, they cannot be made to look cute and careful detailing is needed for identification. Sometimes they can be arranged decoratively, but despite the popularity of fishing (Izaak Walton's classic *Compleat Angler* was first published in 1653) and continuing obligations to eat fish on fast days, a Latin quarto devoted to them was not going to be a bestseller. Many plates bear the names of individual Fellows of the Royal Society who public-spiritedly paid for them to be engraved; some such patronage was essential if the many large illustrations required were to be included in a taxonomic work like this.

Ray also wrote a *Flora of Cambridgeshire*. Baldner and Johnson had studied their respective regions, and local natural history now became a popular subject, generally allied with history and geography. Far away, such work might also be done for their own localities by factors or colonists in Asia or America. Maps of counties as well as of nations and of the world were being printed and collected, and societies and academies were investigating natural resources, industries and curiosities. Natural history included the mineral kingdom as well as animals and plants, all of economic as well as scientific importance, and fossils (or formed stones) began to attract the attention of the curious. That word, coming from the Latin *fossa* for a hole or ditch, had previously meant anything dug up, but by Ray's time its use was becoming restricted to stones resembling living forms. Hilda, the seventh-century abbess of Whitby, was credited with turning overabundant local snakes to stone, in the form of the many ammonites found there; but such stories no longer satisfied more sceptical generations. Ammonites resemble the curling shells of snails in the cabinets of the curious, as illustrated by Lister's daughters; but some are enormous, far larger than any such creature known today: such snails seemed incredible. Big fossil bones might have come from documented giants, like Goliath or Gog and Magog, but otherwise the idea that there were or had been such huge living creatures seemed as preposterous as the story of St Hilda and her miracle. Moreover, the fossils were solid rock and sometimes glittered with iron pyrites, fool's gold. There was no plausible mechanism for an alchemy whereby a dead organism could be exactly transformed into a lump of stone. To many it seemed much less extravagant to suppose that fossils were like crystals, or the basalt hexagons that form the Giant's Causeway, the product of some ordering force or shaping spirit of nature. Debate was by no means confined to England.

The Würzburg professor Johann Beringer was an enthusiastic collector of fossils, publishing illustrations of his striking finds. Unfortunately he was the victim of a famous hoax: two jealous colleagues, taking advantage of his zeal and credulity, began making fossils out of limestone and planting them for him to find. On some they engraved the name of God in Hebrew, as of an artist signing his work. Their victim was delighted, and when chisel marks were pointed out to him chose to perceive them as evidence of divine handiwork and caprice. When the hoax came to light, the chastened Beringer tried to buy up all the copies of the book in which he described the fakes, *Lithographiæ Wirceburgensis* (1726): the perpetrators were disgraced, but some of what came to be called his *Lügensteine*, lying stones, found their way to Oxford University's natural history museum. Moderns should not feel too superior: a whole generation in the twentieth century accepted the 'Piltdown man' as a genuine discovery when it was, like Beringer's fossils, a hoax.

Among those who, like Hooke and Beringer, pondered the status of fossils was Robert Plot (1640–96), professor of chemistry at Oxford and curator of the handsome museum opened in 1683 to house Ashmole's collection (given to the university in 1679). He began a new tradition of publishing the natural, civil and ecclesiastical history of counties or other limited areas. A man of formidable erudition, Plot had studied surveying and was keenly interested in machines and antiquities as well as pharmaceutical chymistry in the tradition of Helmont. He began research for his *Natural History of Oxfordshire* in 1674 and published it in 1676. Elected FRS in 1677, he energetically edited the *Philosophical Transactions*; to stimulate others to join his project he published *Enquiries* with headings and notes to help them describe their native county, and in 1686 brought out the *Natural History of Staffordshire*. There was no pattern for a scientific career; one had to make one's own way. Thus, disappointed of advancement under the new government of William and Mary after 1688, Plot resigned his Oxford posts, married and moved to London in 1690, where he was appointed to a position at the College of Heralds and began work on his book on Middlesex. He died before he could complete it, but his very successful *Oxfordshire*, reprinted four times by Oxford University Press, was edited by his stepson for a new edition in 1705.

The secret was to harness local pride and the support of the landed gentry. *Oxfordshire* begins with a fold-out map of the county surrounded with the gentry's coats of arms and a dedication to the bishop; it has a scale in Oxford miles 'about sixty to a degree' – so long, geographical (or nautical) miles of 1.85 km. The book begins with the elements: heavens and air, waters, earths and stones, including 'formed stones'. Then come plants and brutes, men and women, arts, and antiquities. The plates are dedicated, in a cartouche at the bottom, to whomever has sponsored them. The fossils were of particular interest to Plot, and seven of the sixteen plates are devoted to them. We find, among belemnites and fossil sea urchins, one that 'has perfectly the Shape of an *Owl's* Head... it is a black flint within' and another in the shape of a human foot. The author performed chemical experiments on these formed stones and declared himself puzzled, asking rhetorically:

> *Whether the Stones we find in the Forms of* Shell-fish, *be* Lapides sui generis, *naturally produced by some extraordinary* plastic virtue, *latent in the Earth or Quarries where they are found? Or, whether they rather owe their Form and Figuration to the* Shells *or the* Fishes *they represent, brought to the* places *where they are now found by a* Deluge, Earth-quake, *or some other such means, and there being filled with* Mud, Clay, *and other petrifying* Juices, *have in tract of time been turned into* Stones, *as we now find them, still retaining the same Shape in the whole, with the same* Lineations, Sutures, Eminencies, Cavities, Orifices, Points, *that they had whilst they were* Shells?

Plot, 'upon mature deliberation', was inclined to agree with Lister that they were simply stones, rather than with Hooke and Ray, who believed that they had been animal, that opinion being 'pressed with far more, and more insuperable Difficulties, than the *former*'.[7]

Noah's flood did not seem to Plot adequate to account for all the phenomena, nor did other reported floods and earthquakes; moreover, many fossils resembled no species now to be found living. We see a cautious inductive mind at work, averse to hypotheses. But Plot was also a man of very wide interests and insatiable curiosity as the rest of the book shows, with its accounts of paintings and buildings (notably the roof construction of Wren's spectacular Sheldonian Theatre), of prominent past and present members of the university, farming methods and country traditions, ghosts, tumuli, and Roman and medieval remains. There was something in it for everyone; and in a curious age Plot was by no means unusual in combining natural history with antiquarian interest, though particularly forceful in promoting it.

Among his successors was William Stukeley. He was a friend of Stephen Hales (1677–1761), parson, plant physiologist and chemist, an important pioneer in the study of gases and respiration. Stukeley had joined the Gentlemen's Society in Spalding and was associated with other short-lived groups of the like-minded as polite society and intellectual conversation became an ideal in the new world of the early English Enlightenment – a development that had impressed Voltaire when exiled there in 1726. To Stukeley the elderly Newton first told the story of the apple; and he duly became a prominent FRS, physician and clergyman; also eminent as an antiquary, he took issue with Inigo Jones over Stonehenge. To Jones, this extraordinary and rugged monument with its standing stones and trilithon arches had looked Tuscan, and he attributed it to the Romans. Stukeley saw it and other stone circles as older, part of a sacred landscape, and in two famous (or notorious) books on Stonehenge (1740) and Avebury (1743) argued that the Druids had built them. His romantic error (the monument is thousands of years older than he imagined) lives on in the revived Druid order.

Another enthusiast for the mysterious Druids but a more formidable scholar was William Borlase (1696–1772), a vicar in Cornwall who became a keen mineralogist. He corresponded with Carl Linnaeus in Sweden, and with Emanuel Mendes da Costa (1717–91), who came from Portugal, lived in London, had a wide circle of acquaintances and managed Borlase's election to the Royal Society in 1750. Borlase contributed stones to the famous grotto built by the poet Alexander Pope.[8] In 1754 Borlase published a substantial volume on the antiquities of Cornwall and, in 1758, a companion, *The Natural History of Cornwall.* The subscribers to this volume, listed at the beginning, included William Pitt the Elder (1708–78) and Horace Walpole (1717–97), as well as numerous county gentlemen and clergy, and men of science at home and abroad. The plan is like Plot's; two-thirds of the book is devoted to mineralogy, as befitted a county rich in tin and other metals as well as in china clay. Behind it all is a theology of nature:

> Bounteous Providence has laid her works before us; she has opened the spacious volume of Nature; 'tis our part to read, compare, and understand. NATURAL HISTORY is the handmaid to Providence, collects into a narrower space what is distributed through the Universe, arranging and disposing the several Fossils, Vegetables, and Animals, so as the mind may more readily examine and distinguish their beauties, investigate their causes, combinations, and effects, and rightly know how to apply them to the calls of private and public life.[9]

Usefulness is exemplified in a section of a mine with a description of how it was worked, and illustrations of a steam engine and other machinery as well as of minerals and fossils (attributed to plastic power or the Deluge), plants and animals, and the houses of those who had sponsored the plates.

Plot and Borlase were interested in everything, but by the mid-eighteenth century sex and reproduction, which had fascinated Aristotle, were coming to preoccupy natural historians all over Europe, trying also to bring order to the vast array of animals, plants and minerals confronting them. Reading Willughby's *Ornithology* and meeting Borlase in Cornwall set Thomas Pennant (1726–98) on a career as a naturalist and traveller, mostly within the British Isles. He was FRS and an antiquary; his *British Zoology* (1766) was a great success, and his published tours, most notably in Scotland, made him a wide reputation: Gilbert White's celebrated *Natural History of Selborne* (1789) was based in part on letters sent to him. He was also part of Linnaeus' worldwide correspondence network.

Cultivators of fruit and flowers over the centuries knew practically about fertilisation but the sexual aspect of plant reproduction did not become a part of formal and general knowledge until Grew and Malpighi demonstrated it. With Harvey's work on mammalian embryology, and Leeuwenhoek's discovery of spermatozoa, although nobody saw a mammalian ovum until the nineteenth century, there was much controversy in the eighteenth about animal reproduction. The process might involve epigenesis, the development of the embryo from undifferentiated matter contributed by both sexes, or preformation, in which a minute organism, pre-existing in the male or the female, was set growing by coition. The idea that lowly creatures were spontaneously generated from lifeless matter had been challenged by the experiments of Francesco Redi (1626–97), a member of the Accademia del Cimento and physician to the grand duke of Tuscany. He reported in 1668 putting meat into a number of jars, and covering half of them with muslin to keep flies away. Maggots duly appeared only in those that had been accessible to the insects: Harvey had been right about all life coming from eggs. Their brief lives, quick reproductive cycles and strange metamorphoses, which Swammerdam had followed in his microscopic observations of their life-cycles, made insects particularly fascinating. René-Antoine Ferchault de Réaumur (1683–1757) studied with Jesuits at Poitiers before going to Paris, where his mathematical ability brought him election to the Académie des Sciences in 1708. A polymath, he investigated the manufacture of iron and steel, tinplate, and ceramics (becoming a great promoter of trade and industry), devised a scale of temperature, studied the digestive systems of birds, and tabulated the growth of insects and how their numbers are maintained and controlled. He demonstrated that corals are not plants, but animals; such apparent zoophytes, creatures on the boundary of plants and animals, greatly interested anybody seeking to construct a scale of nature, or chain of being from the simplest creatures up to us. Réaumur was elected to the Royal Society in 1738, and later to the Swedish Academy; and in 1734–42 he published *Mémoire pour server à l'histoire des insectes* in six illustrated volumes, containing much close observation.

In Geneva the young Charles Bonnet (1720–93) read Réaumur and Pluche, and was drawn into natural history. Trained in law, and coming from a Protestant family that had fled from France, he was a bookish boy who became fascinated by insects. He observed the respiration of caterpillars through pores, spiracles; and, astonishingly, that aphids, greenfly, reproduce by virgin birth, parthenogenesis. This convinced him, and many of his contemporaries, of ovism, the idea that the complete animal is contained in miniature in the egg, and that the male merely triggers its development – something evidently not necessary for aphids. Elected to the Académie des Sciences in 1741, the Royal Society in 1743, and subsequently to the Swedish and Danish Academies, he also worked on 'germs' and on mimosa, the 'sensitive plant' that recoils from the touch. In 1762 he published *Considérations sur les corps organisées*, against epigenesis; and, in 1764–5, *Contemplation de la nature*, celebrating the old idea of the scale of being and extending it so that it stretched without gaps from germs to humans. This book was widely translated, accessible and influential. As his eyesight failed, Bonnet described hallucinations that can befall the visually impaired; found usually among the elderly, this syndrome is named after him. He turned to philosophy, arguing for human immortality in a progressive universe.

Bonnet's cousin Abraham Trembley (1710–84) left Geneva when his father was exiled in 1733, and went to the Netherlands to be tutor to Count William Bentinck's (1707–74) two sons. There he took up natural history, investigating parthenogenesis in the wake of Réaumur and Bonnet; and then found himself fascinated by a little polyp that he called hydra after the fabulous many-armed monster that Hercules killed. It had previously been described by Leeuwenhoek, and was generally supposed to be a plant; but Trembley noted that it seemed to wave its tentacles about like an animal. Thinking that animals are killed when cut in half but plants are not, he tried the experiment – and found to his astonishment that the two halves regenerated completely. Moreover, parts of hydras could be grafted together, like plants, and the creatures reproduced by budding. They were true zoophytes, a bridge linking the plant and animal kingdoms in the scale of nature. He published a book about them, *Mémoires*, in 1744; but already the news of this exciting research had led to his being elected to the Royal Society in 1743, and being awarded its Copley Medal. His work, showing how part of a simple organism could transform itself into a whole creature, was taken to be a blow against the preformation favoured by Bonnet.

Another Swiss, Albrecht von Haller (1708–77) from Bern, who studied medicine in Tübingen and Leiden, took over from Boerhaave the role of the pre-eminent medical teacher in Europe when he was appointed a professor at the new University of Göttingen in 1736. He made collections of plants from his beloved Alps, and came by 1753 to feel so homesick in north Germany that he resigned and returned to relative obscurity in Bern. His most important studies had been connected with 'irritability', reaction to stimuli, and the connection of nerves and muscles, making him a founder of neurology; but he became also a prominent supporter of ovism. Sex and reproduction were thus mainstream and controversial topics among zoologists; and it will not surprise us that, when the eighteenth century's greatest systematiser set out to devise a scheme better than that of Joseph Pitton de Tournefort (1656–1708) but just as convenient, he should take up the idea of plant sexuality as the key.

Linnaeus' own career had been launched when, after graduating in medicine from Lund and Uppsala, he had made a tour of Lapland in 1732: Sweden was no longer the great power in the Baltic that it had been, and the war-impoverished government was taking an interest in the remote north. Linnaeus described unfamiliar plants growing there, and his interests shifted from medicine towards botany. He then secured funds to complete his medical education in the Netherlands from 1735 to 1738, being attracted to Boerhaave, though, like him, he took his MD at Harderwijk (because it was cheaper and speedier than Leiden). A minister's son, he had earlier intended, again like Boerhaave and also William Hunter, to become a clergyman, and a theology of nature remained one of his major concerns. After collecting his MD and visiting Leiden to work with Boerhaave and use the magnificent botanic garden there, Linnaeus found a new patron, George Clifford (1685–1760), a director of the VOC (Dutch United East India Company) and wealthy banker, describing his garden in a sumptuous publication – and inducing a banana to bear fruit in his greenhouse. Confronted with all the plants coming in with the Dutch fleets, he displayed his talent for classification, taking sexual reproduction in plants seriously, and seeking, there and back home in Sweden after a visit to England, to pick up where Ray and particularly his French contemporary Tournefort had left off. Whereas Ray had aimed at a natural method, Tournefort's system was artificial, being based on the form of the corolla, the flower; but that meant that it was speedier and more convenient to learn and to apply. Very clear about genera and species, it became widely popular, and Linnaeus used it until he devised a better system.

In the Netherlands he had taken up the task of recording and classifying not only plants but also animals and minerals. The word 'biology', and its associated science of life, was invented about 1800, and earlier natural history was concerned equally with the animal, vegetable and mineral kingdoms. There was much confusion about names, with the same plant being given different names by naturalists: Linnaeus determined upon using Latin, and giving concise descriptions. He hit upon his greatest idea when devising an index: he gave double-barrelled names, binomials, to every animal and plant. The first name denotes the genus, and the second (or trivial) name the species. For plants, while recognising that the true natural system (as in the hands of Ray) would require all aspects of a plant to be considered, Linnaeus used an artificial system based upon counting the sexual parts of the flowers that was much quicker and easier, and usually seemed pretty close to nature.

Plants then fell into groups, like *Diandria* with two stamens, and *Digynia* with two styles or sessile stigmas; and the science could be lightened by jokes about two husbands and two wives, for example, in one bed. The system, extended less happily into the animal and mineral realms, began to catch on: it brought the order especially required in an epoch of ocean voyages when new species were daily being brought back to Europe, or found in remote parts of it like Lapland. In 1735 Linnaeus published *Systema Naturae*, and in 1753 *Species Plantarum*: the latter for plants, and the tenth edition (1758) of *Systema Naturae* for animals, became the accepted starting places for nomenclature thereafter. Particularly striking for the modern reader looking at the animals is the entry under *Homo sapiens*, 'nosce te ipsum' – know thyself. This is followed by a fuller description in which humans were divided into varieties, for all of us belonged to one species; but we shared our genus with *Homo sylvestris*, the orang-utan or wild man of the woods. This makes us much closer to apes than in modern classifications, but was not a matter of great concern at a time when there was general agreement that species were fixed and real entities, forming a ladder or chain. The ladder was not for going up and down, but was more akin to library shelves on which each creature had its place. The orang-utan one level below us was no threat because there was no idea of common ancestry or of evolutionary change: the word 'evolution' appeared in Johnson's *Dictionary* in 1773, but only as meaning the act of unrolling or unfolding. Linnaeus allowed for some hybridisation and possibly for some degeneration, but his schema was static. He stimulated his students, or disciples, to travel the world collecting new species, 'nondescripts', for him to fit into his capacious tables and preserve in his herbarium. Whereas Newton's science was the model for the mathematical, explanatory sciences of the eighteenth century, Linnaeus provided a complementary model for the descriptive sciences, aiming at classifying the vast variety of things, and thus also in its way revealing God's mind and plan. It is hard to say which was the more important.

If the plant was in flower, counting the sexual parts was quick, and it could be placed, for example, among the *monandria trigynia*. A naval officer, colonial official, merchant or ship's surgeon could quickly learn how it was done and get the specimens collected on travels or a voyage into order, ready to be formally described and named back home by an expert, perhaps Linnaeus himself. If it was a nondescript, all the better; but it was interesting, too, to know when a species found in Europe was also shown to occur far away. If the season was wrong and the plant not flowering, it was trickier because a seedling should be cosseted until it did, which was awkward on board ship; otherwise the seeds were brought back to a botanic garden. For flowerless plants, the system did not work so neatly. Linnaeus' binomial names have stuck and no pre-Linnaean name nowadays counts as valid. Linnaeus' static system, in which species were definite and unchanging, was suitable for a hierarchical society, where there was a proper place for everyone and everything. When his disciples, or other travellers, sent home their collections, Linnaeus himself, back in Uppsala, would classify and name them. These became the 'type' species, and ultimately if in doubt about a plant one compared it with the type to see if it was the same. In the field it was good to collect a number of specimens because certified duplicates of types could then be exchanged between major collections to facilitate identifications. Whoever controls such a collection is a powerful figure indeed: Linnaeus' was bought after his death by James Edward Smith (1759–1828), a banker, and brought to London, where it is still housed in the Linnean Society, founded by Smith to promote natural history. Like the relics of saints, type specimens became the focus of a kind of pilgrimage.

In Uppsala in 1728/9 the young Linnaeus had attended the zoology lectures of Olof Rudbeck the Younger (1680–1740), who after studying in Utrecht travelled in 1695 to Lapland. In 1730 Linnaeus lived with Rudbeck and tutored one of his sons; and was stimulated to make his own visit to Lapland. Rudbeck was a skilled painter of birds and used his pictures in his lectures. On the occasion of one of these, on the great crested grebe, Linnaeus noted:

> On his pate he hath a horn of feathers, black, like to an upturned fish-tail. Hath eek a great blackish collar about his neck, yellow on the head, with a white blaze. The Professor discharged 7 shot at him, for as soon as he got sight of the powder he made for the bottom: at last he set his hat before his shot-gun to hide it, and so fired. Has 3–4 speckled eggs. The flesh is evil-tasting.[10]

The natural historians' and collectors' maxim 'what's hit is history, but what's missed is mystery' could be difficult to follow in practice: animals and birds are camouflaged and elusive, and shooting a wary creature is an uncertain business – hunters often come home with nothing bagged. Linnaeus thought Rudbeck's paintings were so good that when he could not himself use the stuffed corpse or the skin of a bird as his type, he used Rudbeck's picture instead. Like almost all his contemporaries, Rudbeck believed that swallows conglobulated and hibernated on the bottom of ponds, where they were seen congregating each autumn, and 'reappear in accordance with the wonderful order of Nature as the trees begin to come into leaf': the hypothesis that such small birds migrated to Africa seemed wildly improbable. Linnaeus' botanical system caught on, especially in Scandinavia, the Netherlands and Britain; but in France the search continued for a natural method that would take more factors into account. Linnaeus' zoological and mineral systems, based as they were on external characters rather than dissection and chemical analysis, proved less satisfactory and had a shorter life. But just as Newton was admired for bringing simplicity, elegance and order into mechanics and astronomy, so Linnaeus came to be respected as the patron of classifying and naming, important not only in natural history but also in chemistry, and in life more generally.

Linnaeus' opening remark about *Homo sapiens*, 'nosce te ipsum', was the Latin version of the inscription from the Temple of Apollo at Delphi: Apollo was the god of light, truth, healing and poetry, of high culture. After this exalted start, Linnaeus next prosaically described the races of mankind before getting to our only congener, *Homo sylvestris*, the orang-utan. The great apes from Java and from Africa were not yet distinguished, but in 1699 Edward Tyson (1651–1708) had published *Orang-Outang, sive Homo Sylvestris*, an anatomical study comparing what we call a chimpanzee with a monkey, a (Barbary) ape and a man. At Oxford, Tyson, the son of a prosperous merchant, studied natural history and anatomy with Plot, in whose presence he dissected a polecat in 1675 as part of an investigation into animal scent glands – chemistry not being the only science that deals in stinks. In 1677 he moved to London, where he made friends with Hooke, becoming FRS in 1679 and a Fellow of the Royal College of Physicians in 1683. He believed in the chain of being, and felt strongly that comparative anatomy was important for medicine. Accordingly, he dissected a porpess (dolphin), which he saw as linking mammals and fish, a female opossum, a rattlesnake and various intestinal worms. The engravings of the dolphin were done from drawings by Hooke, who was present at the dissection, and those of the young chimpanzee by William Cowper (1665–1710), an eminent surgeon whose election to the Royal Society Tyson had promoted in 1698. Among Cowper's various works, his anatomical plates, moving on from the classical style of Vesalius towards greater realism, were greatly admired, and useful for artists as well as doctors. Thomas Huxley used illustrations from Tyson in his famous *Man's Place in Nature* (1863), and examined the skeleton of the chimpanzee he had dissected, now preserved at the Natural History Museum in London.

Tyson explained his enthusiasm for comparative anatomy:

By viewing the same Parts ... together, we may the better observe *Nature's Gradation* in the Formation of *Animal* bodies, and the Transitions made from one to another; than which, nothing can more conduce to the Attainment of the true Knowledge, both of the *Fabrick,* and the *Uses* of the Parts. By following *Nature's* Clew in this wonderful *Labyrinth* of the *Creation*, we may be more easily admitted into her *Secret Recesses*, which Thread if we miss, we must needs err and be bewilder'd.

Systematically dissecting what he called a 'pygmie', increasingly convinced that explorers' references to them must really be to great apes, he was constantly struck by the resemblances to humans; and the most famous illustrations, showing the chimpanzee from the front and back, are striking fold-out plates where he is standing up. In one, he holds a walking stick; in the other, he reaches up to a rope, showing his handlike foot. Tyson noted:

> I have represented him with the *Fingers* of one Hand *bended*, as if kneeling upon his Knuckles, to shew the Action, when he goes on all four: For the Palms of his Hands never touch the Ground, but when he *walks* as a *Quadruped*, 'tis only upon his *Knuckles*. The other Hand is holding a Rope, to shew his climbing; for he will nimbly run up the Tackle of a Ship, or climb a Tree.[11]

Tyson not only compared the creature to men and monkeys, but also appended a philological essay scrutinising ancient references (revived by more recent travellers, both fictional like Mandeville and real) to satyrs, wild men, men with dog's heads and pygmies, and inferring that apes and monkeys were meant.

Huxley's generation saw apes as our cousins, sharing our evolutionary history; but despite the great interest in antiquities that went with the Scientific Revolution, everyone then took it for granted that animals and plants were as they had always been. The great chain had a link in it that corresponded to each creature, and a task for the natural historian was to find missing links, like the opossum and the orang-utan: such links were to be sought in the present, not in the past among fossils. Though Shakespeare's Henry IV had longed to see

> the revolution of the times
> Make mountains level, and the continent, –

Weary of solid firmness, – melt itself
Into the sea![12]

and Copernicus and his successors had revealed the immense, even terrifying, depths of space, everyone took it for granted that only a few thousand years had passed since the Creation: nobody yet supposed that there
was a corresponding abyss of deep time, or saw any need for it. Antiquarians,
digging up barrow tombs, speculating about Druids and writing local
history, found that the biblical chronology gave them plenty of time. And
while Walter Raleigh used his imprisonment in the Tower of London to
write a history of the world, some of the greatest historians of the day, like
Edward Clarendon (1609–74), Gilbert Burnet (1643–1715) and Voltaire,
wrote about their own or recent times, as Thucydides (c. 460–c. 400 BC)
had done in ancient Athens. Given the classical emphasis in education, the
statesmen of Greece and Rome seemed like contemporaries; and the
importance of history seemed to lie in the providing of examples to be
followed or eschewed. Anyone writing about geographical discovery, like
Richard Hakluyt (c.1552–1616), whose manuscript 'Discourse of Western
Planting' (1584) and published *Principall Navigations of the English Nation*
(1589) promoted voyaging and settlement, had to take account of technical
changes making ocean crossings more and more practicable; but the idea
that civil history was primarily concerned with change, perhaps progressive,
was novel.

For Samuel Johnson, history was 'a narration of events and facts delivered with dignity', or maybe just a 'narration', 'knowledge of facts and
events', or a 'picture representing some memorable event'. Natural history
and civil history, both concerned with facts and events, could therefore be
perceived as much less distinct than they appear today. The past was like
the present. God could be seen in the Bible to be working through history,
but we were as sinful as our predecessors, in the same predicaments as they,
and expected to learn from their example. Gibbon had written of the
decline and fall of the Roman Empire, and the idea of subsequent progressive change over time, after the supposed long lapse of the 'Middle Ages',
was a feature of the later eighteenth century, the Enlightenment. Johnson's
contemporary Giambattista Vico (1668–1744), living in Naples and

hedged in by both the Roman and Spanish Inquisitions, proposed in his *Scienza nuova* (1725) a more dynamic vision of history. This new science (the title echoes Galileo) would portray the ancient world as radically different from the modern, because people are members of different societies characterised by different values. Imagination was needed to appreciate behaviour and the motives that lay behind it, in a heroic Homeric world, a city-state based on slavery, a mighty empire or a modern mercantile nation. That idea gradually caught on for human history; and by the middle of the nineteenth century, natural history as envisaged by Darwin and Huxley was indeed a dynamic evolutionary development, through almost unimaginably different epochs to the present time.

It is curious that just at that point natural history began to be perceived as an amateur activity in a newly professional world of serious science, done in laboratories, observatories and museums. But that was not how it was in Linnaeus' day: natural history was as important as natural philosophy and complementary to it. As well as leading to wonder at the riches of the Creation, it promised useful fruits – literally, and through new minerals and ores; plants for eating, for medicine and for decoration; and domestic animals and birds like the alpaca and the turkey. It was also, with astronomy, the 'big science' of the day, involving public money for expensive voyages or overland expeditions, numerous participants, botanic gardens, menageries and museums. With astronomy, it was involved in the measuring and mapping of the world, building upon the work of the pioneers since the 1490s, and making navigation, trade and settlement easier and more profitable. It is to this project that we now turn.

CHAPTER 11

A Global Perspective
Exploring and Measuring

THE UNIVERSE OF GALILEO and Newton was vast and indefinite. Some, like Bentley, thought it infinite but Kepler had argued that an infinite number of stars stretching away to infinity would give us a bright rather than a dark night sky. Stars displayed no parallax and until the nineteenth century their actual distances were unknown; but knowing how big the Earth and the Solar System were did seem possible. Science is concerned with the knowable, and exact measurements are valued both for their own sake and as a rigorous test for ideas. This was already the case in the eighteenth century. By 1700 there were figures for the diameter of the Earth and its distance from the Moon and the Sun, but accurate measurement entailed expanding science and state support for its inquiries. Expensive equipment in observatories was needed, and ships to take observers on prolonged voyages to distant places whose exact position had to be found and charted. Determining longitude on land was difficult, but at sea remained impossible, so navigators continued to estimate it by dead reckoning. To find the ship's speed, they cast a log of wood attached to a line overboard, and saw as the ship sailed on past it how many knots in the cord had to be paid out in a given time: if the time was to be half a minute these knots on the cord were tied 50 feet (15 m) apart. The number of knots released indicated the ship's speed in nautical miles per hour (therefore called 'knots'). Carefully noting the ship's direction, navigators could then plot their course on a chart, dead reckoning being:

[t]he judgement or estimation which is made of the place where a ship is situated, without any observation of the heavenly bodies. It is discovered by keeping an account of the distance she has run by the *log*, and of her course steered by the *compass*; and by rectifying these data by the usual allowances for *drift, lee-way*, &c. according to the ship's known trim. This reckoning, however, is always to be corrected, as often as any good observation of the sun can be obtained.[1]

Such an observation would give the latitude, but the longitude remained uncertain. To be sure of finding landfalls charted by previous navigators, the best practice was to get to the right latitude and run along it until one encountered the coast or island sought. Ocean currents were suspected but largely unknown, and certainly confused dead reckoning. Small islands in oceans might have been plotted 50 miles (80 km) or more from where we now know they are, and they remained elusive to later sailors and often still remain so to historians trying to reconstruct a particular voyage. Once on land, astronomical observations, notably of an eclipse, might fix positions accurately. The term 'log' was extended to mean the record of the ship's course and any happenings on the voyage, kept by the captain; a valuable if laconic source for following the scientific voyagers of the eighteenth century.

The idea of the topsy-turvy world of the Antipodes had engaged satirists in Antiquity and in the Renaissance, but with the Portuguese voyages it began to become a reality: there really were inhabited places in the southern hemisphere. But there was also a great deal of ocean; and, because cosmographers believed that for the world to be stable there must be equal amounts of land north and south of the equator, an enormous unknown southern continent, Terra Australis Incognita, appeared on maps from that of Abraham Ortelius of Antwerp (1527–98) in his famous atlas of 1570 onwards. Magellan had sailed through the strait that now bears his name, and Tierra del Fuego to the south seemed a part of this great landmass; but the discovery of Cape Horn with open ocean to its south disproved that. Most of our Australia was missed by early navigators, but parts of the west and north coasts, Van Diemen's Land (our Tasmania) and New Zealand were known to Abel Tasman of the VOC from his voyages of 1642 and

1644. Compared to the spice islands of Indonesia, they were unpromising and unfriendly, and the Honourable Company was not interested. The unknown continent remained a tantalising mystery until the latter part of the eighteenth century.

In Ortelius' map, the outlines of the continents are otherwise recognisable, but the Americas particularly are not accurate. Later, as his successors tried to keep up to date with information, and trading nations tried to keep their secrets, California was for many years depicted as an island, and various straits leading from western North America into a North-West Passage were confidently shown. Despite some information from travellers, the interiors of the continents of America, Africa and Asia were almost unknown until well into the nineteenth century: indeed, when I had the good fortune to join a student work camp in Madagascar in 1959, parts of that vast island still lacked air or ground survey, and were blank on the map. With improved instruments, surveying on land by triangulation became steadily more accurate, and the mapping of western Europe came to look more modern. To indicate high land on maps, pictures of hills or mountains were slowly replaced by hachuring: at first rudimentary, with lines from the top indicating the course water would take, flowing freely; but systematic from the early eighteenth century, when the lines were drawn broader or closer to indicate relief. About the same time sailors began using contour lines in sea charts, indicating depths, but they were not used for heights until much later. Measurements of height were tricky, for Pascal's method of carrying a barometer involved bulky and fragile apparatus, and survey instruments were insufficiently precise. In unknown territories, travellers would keep itineraries and try to work out just where they had been like dead-reckoning sailors.

In Latin America, the Spanish and Portuguese authorities were as interested as anyone in the natural history and resources of their vast territories, but did their best to damp the curiosity and cupidity of their rivals (especially the Dutch and British, whom they saw as pirates and heretics) in order to prevent them trading with these colonies and thus breaking the Iberian monopoly. When the rivals did get there, they found that the local colonists were, in defiance of their governors, extremely keen to trade, making good, if illegal, bargains. Much of the geography and natural history of Central and South America was therefore, because governments and illegal traders both sought secrecy, generally unknown until the nineteenth century, when Alexander von Humboldt (1769–1859) and his successors were able to explore them and publish their results. Information had been collected under Spanish rule but kept as state secrets, and some of this Spanish science, with splendid illustrations by local artists, is at last being published in our day. Spain's rivals were less secretive, though less successful. The Dutch planted a short-lived colony in northern Brazil, and this generated more public knowledge of birds, animals and plants; but Walter Raleigh's expedition to Guiana in search of the riches of Eldorado was unfruitful, and led to his execution when he returned empty-handed. Britons did settle there subsequently; and Scots colonists were sent to Central America, to Darien, in a disastrous enterprise that bankrupted their country and led to its full union with England in 1707. North of Mexico, Raleigh's Roanoke colony, the original Virginia, had also failed; on the ship's eventual return from England, the settlers were nowhere to be found and presumed dead. Planting settlements far from home was not straightforward. But the colonies in Canada (New France) and New England and on the Atlantic coast succeeded, and maps as well as plants, decorative and useful, were in demand as the colonists prospered, expanding westwards, their population doubling in twenty-five years. The importance of maps for armies was clear as the British and French fought for supremacy, there and elsewhere; so in 1720 the Ecole d'Artillerie was founded in France to teach gunners to survey, the model being copied in the Royal Military Academy at Woolwich near London in 1741. In Britain the Ordnance Survey, set up for artillery, still publishes the standard maps.

Trade with India, Malaysia and China, and then also with colonies in the Americas, meant that there was urgent need for good maps and charts to reduce the number of shipwrecks and the deaths from scurvy that resulted from overextended voyages and ships caught in the doldrums. Pirates, buccaneers and privateers also threatened trade; and the Scientific Revolution was an era of frequent conflict. In distant ports there was in effect a truce, but, coming home from long voyages, mariners might find that there was another war on when they were unexpectedly bombarded by a frigate. Merchants needed to safeguard themselves against losses, and from the thirteenth century the practice of insurance that began in Italy became increasingly general, as Antwerp, and then Amsterdam and Hamburg, came to dominate seaborne trade within Europe and beyond it. Merchants also needed credit to buy and sell goods abroad. Trade was a complicated business: it might happen, for instance, that in South America a voyage that began in a Dutch ship would conclude in one flying, legally or illegally, a Portuguese flag. At first, this system of credit depended upon families, fellow countrymen or co-religionists who could be trusted not to swindle one another; but gradually, as the world became more individual-istic, contracts that could be legally enforced became more prominent. In 1607 the Amsterdam Stock Exchange was founded; and especially after William of Orange's successful landing in England in 1688, Dutch finan-cial institutions, the central bank, insurance and finance across frontiers, were taken up in the City of London. There, brokers met in Lloyd's coffee house, making eighteenth-century London a headquarters of marine insur-ance just as it was of instrument-making. These developments, along with aristocratic passions for horse-racing and gambling, made the study of probability an important part of applied mathematics. Credit for merchants was a relatively short-term, high-risk business, and probably led to a different ethos from later German banking, which grew in the nineteenth century by making long-term investments in industry.

Improved instruments meant that an important point in theoretical physics could be settled. Clearly the Earth, with its mountains, is not a perfect sphere; but these are but pimples, whereas different conceptions of gravity held in the early eighteenth century led to differing predictions that the Earth might be somewhat distorted, a spheroid rather than a sphere. In Britain and the Netherlands, Newton's theory of gravity had been generally accepted by 1727 when he died, although it involved the mysterious attrac-tion between the Earth, the Moon, other planets and the Sun across void space. Because the Earth was spinning rapidly in empty space, he predicted that it would be slightly oblate, flattened at the poles like a satsuma. In France and much of continental Europe opinion was different: thus Fontenelle in his *Eloge* of Newton for the Académie des Sciences, after praising his 'sublime Geometry', wrote:

The continual use of the word Attraction supported by great authority, and perhaps too by the inclination which Sir Isaac is thought to have had for the thing itself, at least makes the Reader familiar with a notion exploded by the Cartesians, and whose condemnation had been ratified by all the rest of the Philosophers; and we must now be upon our guard, lest we imagine that there is any reality in it, and so expose our selves to the danger of believing that we comprehend it.[2]

Although he admired the way that Newton could 'foretel events', notably that when Saturn and Jupiter were close to each other their orbits would be disturbed by their mutual attraction, he believed that real science demanded a plausible explanation as well as a mathematical law. He, and others in the Académie, believed in an æther like Descartes' in which planets were immersed and which kept them in motion. If the Earth were being swirled along, hugged and spun by this æther, then the equator would be squeezed and the planet would be a prolate spheroid, slightly elongated like a plum. In France and elsewhere in temperate Europe, the distance or arc that you had to go to get one degree further north had been carefully measured by surveyors. If the Earth were a sphere, this distance would be the same everywhere; if a spheroid, then further measurements made near the poles and the equator would give different results, and show if it was prolate or oblate. That was a project for the Académie.

Jacques Cassini (1677–1756), who inherited his father's post of director of the Paris observatory and his cartographic responsibilities, emerged as the staunch defender of the prolate Earth, while Pierre-Louis Maupertuis was the Newtonians' champion. The latter had sought, with Alexis Claude Clairaut (1713–65) and Pierre Bouguer (1698–1758), to turn Newton's geometrical presentation of his physics into algebraic form, and then extend it using Leibniz's version of the differential and integral calculuses: such mathematical analysis proved a flexible and powerful tool, neglected in Britain under Newton's shadow and foreign also to Cassini. To settle the question, two expeditions were mounted: Bouguer, accompanied by the attractive and dashing polymath Charles-Marie de La Condamine (1701–74) and the botanist Joseph de Jussieu (1704–79), set off for Peru in 1735. French power ensured that they had the support of the Spanish government. Such voyages were 'big science', promoted by the Académie des Sciences, requiring ships, a well-equipped team and much time: La Condamine was away for ten years, and Jussieu for thirty-six. Jussieu sent plants and seeds to other members of his distinguished family at the Jardin du Roi; the physicists quarrelled badly but, working their way southwards from Quito (now in Ecuador), made the observations with their sensitive instruments that led them to conclude (as they had expected) that Newton was right. They also found curious discrepancies which La Condamine explained: Newtonian gravity is not an Aristotelian pull towards the centre of the Earth, but mutual attraction of particles of matter. Among the high Andes, therefore, the bob on the plumb line used to level the instruments was drawn not only downwards but also sideways towards the mountains. Once the readings had been corrected, the deviation provided a way of comparing the mass of the Earth and that of a mountain; and since the mass of the latter could be reasonably estimated (or guessed at), so could the former. Jussieu was left behind in South America, Bouguer died in 1758, and La Condamine, who had published his *Journal de voyage* (including an account of his daring journey home down the Amazon) in 1751, outlived the other participants. The credit for this expedition's success therefore went to him rather than to his less personable comrades: he had weighed the Earth.

The delays in getting to South America and setting up there meant that La Condamine and company's results were hardly hot news: they had been anticipated by the polar expedition of Maupertuis and Clairaut, who went to the head of the Baltic, recruiting Anders Celsius (1701–44) from Sweden on the way. Using instruments made by George Graham, they measured their arc in winter at the mouth of the frozen river Tornio, just below the Arctic Circle. The region is thickly wooded with birch and spruce, but on the ice they could get clear views and readings. The outcome was that the Earth was found to be slightly flattened at the pole, confirming Newton's view. In 1740 Maupertuis had his portrait painted by Robert Levrac-Tournières (1667–1752): he is clad in a Lapland costume, his left hand (palm downwards) on a globe as if flattening it, and his right rather uncomfortably gesturing above a sheet of geometrical constructions out beyond the picture into the light. His triumph was incomplete: he was mocked by Voltaire, fell out with colleagues in the Académie in the gossipy,

salon-dominated Parisian world of individual and institutional competition, and in 1745 accepted Frederick the Great's invitation to head his new
Prussian Academy. There, among other activities, he studied a family in
which various members across several generations had been born with a
sixth finger, a pioneering work of what we now call genetics.

The development of Newtonian physics thus became an Anglo-French
collaboration, but in the political sphere the struggle between the nations
continued in a series of wars, with Spain allied to France. In 1739 George
Anson (1697–1762) was put in command of a squadron ordered to sail into
the Pacific, harry the Spaniards there, and capture the treasure galleon that
sailed annually between Acapulco in Mexico and Manila in the Philippines.
He set off with six ships; it was a desperate voyage and, their crews weakened by scurvy and faced with storms, only three of the ships managed to
round Cape Horn. They captured various trading vessels, sailing up to
Acapulco, where they charted the harbour, but found no galleon. Reduced
to just two ships, the depleted squadron crossed the Pacific, putting at last
into Tinian, where fresh meat and fruit revived the sickly survivors; by now
they were all in one vessel, the *Centurion*, its rotting companion, the
Gloucester, having been evacuated and blown up in mid-ocean, making a
spectacular mushroom cloud. Urgent refitting was done in Macao, to the
discomfort of British merchants there, disconcerted by a warship's threatening presence, and under a watchful and obstructive Chinese bureaucracy.
The ship then sailed back to the Philippines, to stooge around near where
the treasure galleon was expected to make her landfall. After a month's
waiting she duly came in sight, and following a ninety-minute engagement
the *Centurion*'s heavier guns prevailed and the supposedly unsinkable
galleon surrendered. Anson returned to England with booty worth about
£400,000, a huge sum, which was duly paraded through London, where he
was fêted and became a national hero. By other measures, the exploit had
been disastrous: from the original squadron only 145 men returned; 1,300
had died from disease (only four by enemy action); and only one ship came
back from the four-year cruise. The account of the voyage, a muted triumph,
ends soberly with the lesson: 'Though prudence, intrepidity, and persever-
ance united, are not exempt from the blows of adverse fortune; yet in a long
series of transactions, they usually rise superior to its power, and in the end

rarely fail of proving successful.'³ Despite its horrors, which stimulated the search for a cure for scurvy, the voyage had opened up the Pacific to British seafarers; and when peace came, Anson was followed by two other naval captains, John Byron (1723–86; the poet's grandfather) in 1764–6, and his lieutenant Philip Carteret (1733–96) in 1766–9, both making a fuller and happier reconnaissance.

Anson had noted the strategic importance of the Falkland Islands – the Islas Malvinas to Spaniards – and Byron's orders were to survey and formally claim them for Britain; and then to search for the southern continent, proceeding into the Pacific. His ship, HMS *Dolphin*, was experimentally sheathed with copper to preserve the hull from worms. This innovation was not only technically but also operationally important, in giving the vessel more speed and manoeuvrability – it also stimulated copper-mining back home. In the event, it worked well, and Byron circumnavigated the globe in record time. Carteret then took command, returning to the Falklands and crossing the Pacific in search of islands: happening upon delightful Tahiti, he and his crew spent nearly six weeks there, and on 25 July 1767 carefully observed an eclipse of the Sun, enabling them to calculate their longitude accurately. George Robertson (c. 1732–c. 1799), master of the ship under Carteret, wrote of this earthly paradise when they left it:

> I realy do believe their was a vast many of the Country people who would have willingly come home with us, if we could have taken them, and their was some of our Men, who said they would stay at this place, if they were sure of a Ship to come home with[in] a few years.⁴

Meanwhile, in 1764, the French had set up a base on the Falklands; but in 1766 Louis de Bougainville (1729–1811) was sent to evacuate it, because the territory had been ceded to Spain. His voyage around the world took three years and also included a stop at Tahiti, whence he brought back a Tahitian to Paris. Comparing his dead reckonings with astronomical observations, he detected long-suspected ocean currents:

> From the Isle of Ta[h]iti, the currents had carried us much to the westward. By this means it might be explained, why all the navigators

who have crossed the Pacific ocean have fallen in with New Guinea much sooner than they ought. They have likewise given this ocean not by far so great an extent from east to west as it really has.[5]

He carried a botanist, but although we delight in the *Bougainvillea* named after him the voyage was not especially profitable for natural history.

The discovery of Tahiti came at an opportune time. Halley had predicted that in 1761 and 1769 the orbit of Venus would carry it across the face of the Sun. Such pairs of transits occur approximately a century apart. Plotting its path from distant but accurately surveyed places on Earth (to give an exact and very long baseline) would enable our distance from Venus and from the Sun to be calculated; expeditions were mounted and men of science from Europe and in America commissioned to observe the event in 1761. In the event the times at which Venus entered and left the Sun's disc proved trickier to observe than had been hoped, and there were too many observations, made with instruments of different precision by observers of variable reliability, for a firm result to be calculated. But the second transit would give the chance to learn from the first; and among the expeditions mounted was one to Tahiti, where Point Venus, dedicated to the planet and the goddess of love, seemed a highly appropriate location for an observatory, almost as far as one could get from Greenwich. The captain chosen to lead the expedition was Lieutenant James Cook, who had experience of wartime survey in Canada, and he picked a Whitby collier, calling her HMS *Endeavour*. Built for the stormy coastal waters of the North Sea, colliers were sturdy and capacious rather than speedy, unlike the warships that had preceded Cook's expedition; he was not to be tempted, like Byron, to hurry home, and was away nearly three years. Cook and his officers were expected to make observations, both of Venus and for latitude and longitude, but he also took Charles Green (1735–71), an astronomer nominated by the Royal Society. Joseph Banks was also on board as a gentleman passenger making a global rather than European Grand Tour. A keen and wealthy botanist with experience of Newfoundland and Labrador, he brought with him Linnaeus' pupil Daniel Solander (1733–82) and Sydney Parkinson (c.1745–71), a talented botanical and zoological artist. The rich visual and cartographic record, the ethnography, and the collections brought home made this a voyage of huge cultural importance.

After enjoying the beauties of Tahiti, and observing Venus, they set sail accompanied by Tupaia, a Tahitian prepared to exile himself under Banks' aegis:

> He is certainly a most proper man, well born cheif *Tahowa* or preist of this Island, consequently skilled in the mysteries of their religion; but what makes him more than anything else desireable is his experience in the navigation of these people and knowledge of the Islands in these seas.[6]

Cook had orders to sail south and search for the elusive continent there. His route took him to Aotearoa, our New Zealand, where to everyone's astonishment the Maoris understood Tahitian; but, unlike the Tahitians, they did not welcome the strangers. On his voyages, Cook and his companions collected vocabularies from the various languages they heard and sometimes, as with Tahitian, learned to speak them.

They circumnavigated and charted the North and South Islands, proving that New Zealand was not a promontory from a continent, and continued westwards until they met the unknown eastern shore of Australia, calling it New South Wales. At Botany Bay, Banks and Solander botanised; then the ship followed the coast northwards, dodging in and out of the Great Barrier Reef where eventually the *Endeavour* was damaged and had to put in to the estuary of what they called the Endeavour river for urgent repairs. There they saw, and Parkinson sketched, a kangaroo. The repairs were temporary, and the ship by now needed an overhaul; Cook therefore headed for the VOC base at Batavia (Jakarta) where the Dutch had a dockyard. Up until then, the crew had been remarkably healthy: in preventing scurvy, Cook kept everyone busy to keep up morale, and took a keen interest in diet, buying fresh meat whenever possible, collecting 'wild cellery' and other greens and using them with 'Sour krout, Portable Soup and Malt' under the 'care and Vigilance of Mr Munkhous the Surgeon'.[7] His famous stratagem to get the men to eat these things was to serve them to the officers first, which made them seem desirable delicacies. Sadly,

Batavia was unhealthy: the crew caught fevers, and many, including Green, Parkinson and Tupaia, died – scientific voyaging always had its martyrs. On the return to England, Banks was lionised and in 1778 became president of the Royal Society, devoting his life to science administration and the promotion of empire in the corridors of power until his death over forty years later. Darwin's role on the *Beagle* in the 1830s was like Banks' on the *Endeavour*, but with no entourage.

Meanwhile, back in 1714, the British Parliament had set up a Board of Longitude and offered a prize, the large sum of £20,000, for determining longitude within defined limits of accuracy. It was to be many years before it could be claimed, by John Harrison (1693–1776). Harrison was by trade a joiner, but took up clockmaking, inventing the gridiron pendulum, weighted with brass and steel bars, which kept time whatever the temperature, and making bearings from lignum vitae, an oily tropical hardwood that needed no lubrication. In 1730 he visited London; Halley introduced him to Graham, who assisted him over the years with money and advice, and in 1735 Harrison made a clock that would keep time at sea. Halley and Graham recommended it to the Admiralty, and it was tried on a voyage to Lisbon when indeed it performed well. But this instrument did not meet the conditions for the award, so Harrison returned to work. In 1749 the Royal Society awarded him the Copley Medal; and when in 1758 he saw accurate pocket watches made by Graham's successor as leading instrument maker Thomas Mudge, he was stimulated to make his chronometer in that miniaturised and convenient form. In 1760 it was ready for trials across the Atlantic and back. It performed well, but there was a problem. The sceptical astronomer royal and sticklers on the Board of Longitude still had to be convinced. Harrison's watch and its successors kept perfect time within the prescribed limits, but each one had its own 'rate of going': it would steadily gain or lose a small amount each week. Only when this aspect of the working of this precision instrument was appreciated and taken into account could it be used to determine longitude: it needed to be 'rated' against a standard clock first. Only then could Harrison's watch be seen to have triumphed: but to get the full prize he was supposed to make another and demonstrate how it was put together. He demurred, but allowed his watch to be copied. After petitioning Parliament, and despite cavils, he got his money in the end, three years before his death.

The astronomer royal from 1765 had been Nevil Maskelyne (1732–1811), who developed a method of finding longitude by accurately observing the distance between the Sun and the Moon, and then consulting tables he computed and published regularly from 1767 as *The Nautical Almanac*. His background was very different from Harrison's: from Westminster School he went to Cambridge, was ordained, got a Fellowship at Trinity College and was sent by the Royal Society to St Helena in 1760–1 to observe the transit of Venus (sadly, it was too cloudy on the day to see it). In 1774 round Schiehallion, an isolated mountain in Scotland, he confirmed La Condamine's observation in the Andes by taking measurements from different sides – another weighing of the Earth in more convenient circumstances. Suspicious of chronometers, if not actually hostile to them, he proved to be extremely strict in demanding that Harrison's exactly meet the conditions set. In the end, Anson's shipmate John Campbell (c. 1720–90) was responsible for the successful trials of the instrument in 1764, and in 1785/6 for the testing of Mudge's version on a voyage to Newfoundland and back. Thenceforth, for captains of ships carrying a chronometer and checking it by lunar observation (using *The Nautical Almanac*), the problem of finding longitude that had beset navigators since the first voyages of discovery was solved. At Greenwich rating chronometers became increasingly part of the duties of the observatory.

On his second voyage, 1772–5, Cook had two ships, the *Resolution* and *Adventure*, and had realised that sailing round the world eastwards was much easier than westwards, because of prevailing winds. Now, to fix his position Cook carried 'watches', three chronometers made by John Arnold (1736–99) that all broke down, and one made by Larcum Kendall (1721–95), copying Harrison, that had cost £450 (a year's salary for a professional man) and worked astonishingly well all through a voyage back and forth between tropical and Antarctic seas. Using it, Cook silently moved New Zealand on his charts to the right place. He also had *The Nautical Almanac* so that he could check the running of the watch whenever he managed to set up an observatory on land. He was to seek the southern continent, recuperating in New Zealand and Tahiti between repeated expeditions into high southern latitudes in the southern summers. Amidst the ice, he got below 71°S at one point, in effect circumnavigating Antarctica without actually seeing it, and proving that if the continent existed it was uninhabitable and relatively small. He wrote: 'I who had ambition not only to go farther than any one had done before, but as far as it was possible for man to go, was not sorry . . . [that] we could not proceed one Inch farther to the South.'[8] This voyage, and Cook's third voyage, which took him up the west coast of North America, seeking in vain for a North-West Passage, to the northern ice fields above 70°N, almost completed the task of mapping the continents inaugurated by Iberian navigators four hundred years earlier. Blocked by the Arctic ice, Cook wrote in July 1779:

> It is now clearly impossible to proceed in the least farther to the N°ward upon this Coast and it is equally as improbable that this amazing mass of Ice should be dissolv'd by the few remaining Summer weeks . . . it will remain as it now is a most insurmountable barrier to every attempt we can possibly make.[9]

Accordingly, he followed the Asiatic coast southwards, to Kamchatka, where there was a Russian settlement. Although later on during this voyage Cook was killed in Hawaii, his tradition lived on in the Royal Navy through the officers and midshipmen he had trained; and surveys for trade, military and scientific purposes, filling in the map in greater detail, continued. Meanwhile, from the many observations of Venus made in 1769 from all around the world, astronomers in national observatories computed our distance from the Sun, coming up with somewhat discordant answers. A firm figure for this crucial astronomical unit was not forthcoming until after the great mathematician Carl Friedrich Gauss (1777–1855) plotted the first bell-shaped 'error curve', showing how observations are scattered about the true value (our Gaussian distribution), and thus indicating how this mass of data could all at last be taken into account. Only when error was understood could truth be found.

Banks, Solander, Parkinson and Green not only exemplified a tradition of travelling with scientific interest or objectives, they reinforced it so that in the nineteenth century many distinguished scientific careers began with a voyage. Apart from the many ship's surgeons and other seamen, merchants and adventurers who had recorded what they saw, there were travellers seeking plants for remedies, for food or to introduce into flower gardens. Linnaeus' disciples into foreign parts seeking more and more new, 'nondescript' plants and animals: it became almost as important to know of a plant as of an island whether it was new or a rediscovery. They were men whose MD degree was done under Linnaeus, who had usually written the dissertation they defended, and whose taxonomic project they were keen to advance. Solander was back working in England under Banks' patronage (as his librarian) on the collections they had made, and some plants had already been engraved on copper, when he died; and Banks, disheartened and by now very busy as president of the Royal Society, let the project drop. Of the other Linnaeans, Pehr Kalm (1716–79) went to North America seeking in particular plants that might do well in Sweden; after making extensive collections, he returned safely and wrote up his travels, becoming a professor at Åbo and an authority on economic botany. Less happy were Christopher Ternström (1703–46), who died on his journey to China, and Frederik Hasselquist (1722–52), who died in Smyrna (Izmir) after making rich collections in the Middle East that eventually by a roundabout route found their way to Linnaeus. Pehr Loefling (1729–56), the favourite pupil, went to Spain and on to South America, dying in what is now Venezuela; again, Linnaeus received his manuscripts and published his journal.

More happily, Anders Sparrman (1748–1820) was picked up by Cook at the Cape of Good Hope on his second voyage. After botanising round the South Pacific, on the homeward journey he left the ship at the Cape to continue his researches into that rich flora. Carl Peter Thunberg (1743–1828), pupil and eventual successor of Linnaeus at Uppsala, went to Paris and the Netherlands to continue his studies, and got the chance to go to Japan with the VOC ship that sailed there annually. The country was tantalisingly closed to foreigners from the West, except for this carefully monitored mercantile link, open to the scientific opportunist at the trading station on Deshima Island in Nagasaki harbour. Engelbert Kaempfer (1651–1716), well educated at a number of German universities, had spent two years (1690–2) there as physician and, despite narrow regulations, had established good relations with the interpreters, and the officers of the island, who daily come over to us ... liberally assisting them, as I did, with my advice and medicines, and with what information I was able to give them in Astronomy and Mathematicks, and with a cordial and plentiful supply of European liquors ... I could freely put to them what questions I pleased.[10]

Among other things, he described the ginkgo tree and brought seeds back.

After his death, Kaempfer's papers were bought by Sloane, who got his librarian to translate his *History of Japan* and published it in 1727. Thunberg, inspired by Linnaeus, wanted to update it, and in order to seem Dutch, so as to be allowed to remain on Deshima, spent three years as a surgeon at the Cape of Good Hope, learning the language and also studying the plants, of which he collected over three thousand, one-third of them new to science. Then, after a stay in Java, he reached Japan in 1775, spending a year there. Any internal travel was greatly restricted but again he made contacts and built up an extensive collection of plants; back home, in 1784, appointed professor at Uppsala, he published *Flora Japonica*, with the plants ordered in the Linnean manner. Linnaeus had died in 1778, but he and his disciples had launched the great project of describing and classifying the plants of the world that would form the basis for understanding their distribution and characteristics, and for naturalising useful plants from one continent in another. This benevolent-seeming enterprise, improving on Creation through scientific exploration and understanding, provided a rationale for imperial expansion in the mind of Banks and other powerful and enlightened natural philosophers. To later eyes, by contrast, it could be perceived as placing populations, faunas and floras at the mercy of Westerners, and in the hands of a science that delivered its benefits chiefly to others.

It was not always like that, with only passive victims. In New Zealand a party from the *Adventure* were killed and eaten by Maoris. Their ship was accompanying Cook on the *Resolution*, but in Antarctic fogs the ships had lost sight of each other and the *Adventure* had headed for this agreed rendezvous. In the event, shocked, the surviving crew members had sailed sadly home without waiting longer for Cook. This was an extreme example of unfortunate cross-cultural contact, but Cook himself was later killed in a misunderstanding on Hawaii. Voyagers even to the old civilisations of India, China and Japan, Mexico and Peru, wondered at, and sometimes fell foul of, their different social systems, which they tried to understand, interpret and exploit much as they had in dealing with Islam in the Mediterranean. Some were just there to make money, soldiers of fortune or soldiers supporting commerce or colonisation; but others were deeply curious. Such explorers brought cultural baggage, displaying various degrees of sympathy and antipathy, using classical and biblical as well as pragmatic analogies, some seeing Utopia where others found despotism. These voyages involved encounters with unfamiliar peoples, nomads and those living chiefly by hunting and gathering, whose resources were limited, and who might react with warmth and interest, but often with suspicion – especially in relation to those who were reluctant to share or overstayed their welcome.

Even in parts of Europe such nomadic lifestyles could be the norm, as Linnaeus noted in Lapland:

Ovid's description of the silver age is still applicable to the native inhabitants of Lapland. Their soil is not wounded by the plough, nor is the iron din of arms to be heard; neither have mankind found their way to the bowels of the earth, nor do they engage in wars to define its boundaries. They perpetually change their abode, live in tents, and follow a pastoral life, just like the patriarchs of old.[11]

The Laplanders in their harsh environment were able to hang on to their way of life; but those who had been living in the Americas and the Caribbean were less fortunate. Boyle, believing that everyone should hear the Christian gospel, had financed a translation of the Bible into a local language in North America; but evangelism was soon followed by bringing the various 'nations' into the Anglo-French wars on one side or the other, to their ultimate dispossession and destruction. Traders, captives and missionaries learned the native population's languages, though with little formal understanding of grammar and structure; and from the Roanoke landings onwards, some curious and sympathetic Westerners sought to understand their social systems, much information being collected and some of it published in travellers' narratives.

As well as ruthless conquistadores in Mexico and Central and South America, there were from the beginning some humane explorers, missionaries and even colonists who recognised and respected their fellows whatever their colour, culture or circumstances, and European captives throughout America who 'went native'; and sailors were never backward in establishing relations with women, though that could make trouble. Bougainville, however, noted ungallantly of the Fuegians:

> These savages are short, ugly, meagre, and have an insupportable stench about them. They are almost naked; having no other dress than wretched seal-skins, too little to wrap themselves in.... Their women are hideous.... They had all of them bad teeth, and, I believe, we must attribute that to their custom of eating shellfish boiling hot, though half raw. Upon the whole, they seem to be a good people; but they are so weak, that one is almost tempted to think the worse of them on that account. We thought we observed that they were superstitious, and believed in evil genii.[12]

The Fuegians posed a continuing problem to those who believed in noble savages, or expected naked humans to resemble the gods and goddesses of ancient Greece and Rome. A man in such circumstances seemed indeed no more than 'a poor bare forked animal',[13] and utter nakedness, there, in Australia in Africa and in the South Seas, continued to appal travellers from Europe. Indeed, I remember the shock of meeting a stark-naked herd boy on a mountain in Uganda in 1959.

By Bougainville and Cook's time, captains were firmly instructed to avoid violence, even when they met with hostility as uninvited guests. That fitted with Cook's and Banks' temperaments; they organised trading across a line in the sand, but sometimes, as when shiny brass equipment was purloined, a firm response (often holding someone hostage until it was returned) was required. As a warning, and display of power, the ship's gun might be brought up and fired; and tempers on one side or the other were sometimes lost amid disappointed expectations and misunderstandings. Enlightened voyagers did their best, following their orders, to understand and respect the cultures and individuals among whom they found themselves, though periodically things ended badly; and soon, to Cook's distress, aggressive Westerners following in his track, infected with venereal disease and unscrupulous about selling weapons and alcohol, spoiled what had seemed to benevolent outsiders an earthly Paradise.

Human diversity generated headaches. If, according to the Bible, we all descended from Adam and indeed Noah, and had been dispersed, divided by different languages, from Babel, how had the ancestors of the Australians, the Fijians and the Peruvians got there? The Maoris knew that their ancestors had arrived in Aotearoa only a few hundred years before Cook got there; but other traditions involved thousands of years as the human race spread from Mesopotamia over vast distances and across oceans. It was an important part of Christian doctrine that all humans had Adam and Eve for their forebears; but careful reading of the Bible, and thinking about diversity, might lead to a different conclusion. Common descent meant sharing in Original Sin, but also in common humanity. But heretics had long asked where Adam and Eve's sons found wives, and in 1655 Isaac La Peyrère (1596–1676) published a book, *Preadamitae*, arguing that there had been humans long before Adam. He was denounced and made to recant; but the problem did not go away, especially as the idea that different races might not all be brothers under the skin was politically useful in the days of slavery. The notion that other races were distinct species, who might therefore be treated as domesticated animals, found increasing favour, notably later in antebellum America but not only there; and would be quantified in nineteenth-century measurements of cranial capacity, as early anthropologists became obsessed with skulls in efforts to distinguish races and establish a scale or hierarchy of human varieties. But for both the Catholic religious orders (Franciscans, Dominicans, Jesuits and others) and the Protestant evangelicals of the eighteenth century, the brotherhood of man was an imperative to convert the heathen. Trading companies saw things differently, and did not want to upset the apple cart: Reginald Heber (1783–1826), bishop of Calcutta, remarked ruefully about native converts: 'nothing is to be got [for them] by turning Christian but the ill-will of their old friends, and, in most instances hitherto, the suspicion and discountenance of their new rulers', the East India Company, which feared that reports of conversions would arouse suspicion and unrest.[14]

Even if human varieties were hard to discriminate, exactly where on Earth the various races dwelt could now be fixed by surveys using exact, beautifully made equipment with vernier scales and telescopic sights, by observations of the Sun and Moon, and by chronometers. Climate was also beginning to be systematically documented, as rain gauges, barometers and recording of weather became commonplace; though portable barometers, with the mercury in a leather bag suitable for travellers and sailors, only date from the end of the eighteenth century. But measuring heat was not at first so easy. Galileo had invented a thermometer, but because his lacked a reproducible scale, when the instrument was broken the observations could not be repeated or compared. Gabriel Daniel Fahrenheit (1686–1736) had, by 1714, made capillary tubes with uniform bores, and constructed two thermometers filled with alcohol which fully agreed in their readings; but he then changed to using mercury and announced his method to the Royal Society in 1724. After trying various scales, he settled on the one that bears his name: zero for a freezing mixture of ice and salt, 32 degrees for water and ice, and 212 degrees for boiling water, which because water can be supercooled and superheated is not wholly unambiguous. Meanwhile, Joseph-Nicolas Delisle (1688–1768), invited to St Petersburg by Peter the Great, had invented a scale running from 150 degrees as the temperature of ice and water to zero for boiling water. Thus Stepan Petrovich Krasheninnikov (1711–55), perhaps the first native Russian botanist, reported on Kamchatka what we would see as sub-zero temperatures: 'The winter is moderate and constant, so that there are neither such severe frosts nor sudden thaws as in *Jakutski*. The mercury in *de l'Isle's* thermometer was between 160 and 180 degrees.'[15] Later, in a table, he used Fahrenheit temperatures as well. In 1730 Réaumur invented a thermometer filled with aqueous alcohol, in which water froze at zero and boiled at 80 degrees; improved versions using mercury became standard over much of Europe. Celsius in 1742 presented to the Swedish Academy his centigrade scale, where water freezes at zero and boils at 100 degrees, and this found favour after the French Revolution when the metre, the kilogram and the litre were introduced. He and others toyed with scales that, like Delisle's, ran to our eyes 'backwards', but these dropped out of use.

Now there was a new way of measuring heights for mapping. Because atmospheric pressure falls as one goes upwards, so does the boiling point of water: as anyone will know who has tried, hungrily cooking rice at about 10,000 feet (3,000 m) is a frustratingly tedious business because the boiling water is cooler than it would be at ground level. The drop could be quantified and, given an accurate thermometer, the climber or surveyor would light a fire, boil some water and see how hot it was. This is easier said than done, given conditions on mountain tops; but it became an accepted way of doing things, a form of indirect measurement. More directly, as we know from making jam, measuring temperatures can be an important point of quality control, ensuring in this case that the jam will set.

Glass thermometers were thus useful and portable (though brittle), but they could not be used for high temperatures because they would melt. The potter Josiah Wedgwood (1730–95) devised a pyrometer based not upon expansion but upon contraction. Uniform china-clay pellets placed in a kiln for a standard time were dropped down a calibrated groove: the further they fell, the higher was the temperature. This gave him a better method of quality control over his firings than relying on the experience and judgement of potters. But although the pellets and grooves could be standardised so that his pyrometers could also be used by metalworkers, there was no way in which his scale could be connected to those of Fahrenheit or Celsius. He used an arbitrary scale, but while temperatures in 'degrees Wedgwood' could be compared with one another, they stood alone. Between the heat of furnaces and of ordinary life there was, until the nineteenth century, a great gulf: the clay does not contract at temperatures below the softening point of glass.

Measurement was and is the key to exact knowledge. Pilgrims in the Holy Land had carefully measured monuments there, sometimes in order to make replicas back home. But for the volatile mystic (and author of the first English autobiography) Margery Kempe (1364–after 1448) and others the very tape used for measurement became a vehicle of grace. In the biblical Wisdom of Solomon comes the famous passage 'by measure and number and weight thou didst order all things', and this God-like activity became one of the prime objectives of modern science.[16] For trade and government, standard units might seem essential, but in fact right through the period of the Scientific Revolution there were all kinds of local weights and measures, defined with variable rigour. Deniers, lines, inches and feet, fathoms and toises, miles and leagues, differed from country to country, and within countries; and so did pounds and ounces, grains, scruples, pennyweights and drams or drachms, pints, gallons, puncheons, chauldrons and hogsheads. Helpful editors noted: 'The diversity in the weights and measures of different nations occasions considerable trouble to the readers of foreign books of science, and sometimes even prevents them from fully comprehending their authors.'[17] They therefore published comparative tables to assist conversions, and in translating books and papers recalculated the figures in the original where possible for the assistance of their readers. Despite all the difficulties, by the time of Cook's death there were reasonably agreed figures for the vast distances within the Solar System, the precise dimensions of the Earth, the shape of continents and the latitude and longitude of islands, the relative proximity of towns and cities (and in some cases their populations), the heights of human beings (the reported 'giants' of Patagonia turned out to be about 6 feet, or 1.8 m) and, through Hooke, Leeuwenhoek and their successors, the sizes of the microscopic creatures they saw through microscopes. Moreover, the ideal of public knowledge, of science freely available for the improvement of the human condition, was widespread by the second half of the eighteenth century. It is to this gathering momentum that we next turn.

CHAPTER 12

ENLIGHTENMENT
LEISURE, ELECTRICITY AND CHEMISTRY

THE EIGHTEENTH-CENTURY ENLIGHTENMENT is thought of as a 'project' to make society more rational, more empirical, more 'scientific'. Certainly, at different speeds in different countries, as they recovered from the wars of religion of the seventeenth century, there came a new optimism. Though the four Apocalyptic Horsemen (War, Famine, Pestilence and Death) remained as active as ever, the hope of progress replaced the expectation of Armageddon. While Bougainville, Cook and their accompanying natural historians were filling in the world map, and preparing the way for the great empires of the nineteenth century through this big-scale science, back at home what we call science and technology were growing more and more important in social, political and economic life, as well as in the genteel intellectual culture that was becoming a feature of the more leisured and peaceful lives of the better-off in Europe and the American colonies. If science is insatiable curiosity, then it was as ever much in evidence and now much encouraged; but even the unimaginative were having to take more note of science as the vision of Bacon, Galileo and Descartes, in which experiment and mathematics would bring understanding and mastery of nature, was being realised. Systematic observation and calculation were ubiquitous. Science, moreover, was momentous: beginning with awe and wonder in the face of beauty and truth, bringing an ethic of plain-speaking and plain-dealing, it was revealing order and harmony, it was cumulative, it was proving to be useful knowledge – and knowledge was proving to be power.

Those of us who were around in the 1960s know how far our mundane experience of that supposedly amazing, notorious decade differs from what seems a consensus among commentators too young to remember it. Like the Swinging Sixties and the Scientific Revolution, the Enlightenment is a historians' category, and as such it is fuzzy and would be foreign to many, including people practising science, who lived through it. Historians differ over whether the Enlightenment was one great movement, or whether there were several distinct versions, famously in France and Scotland, but also in England, Germany and elsewhere, with particular national charac-teristics (although there were as yet no nation-states in the modern sense). They also distinguish a moderate Enlightenment from the radical Enlightenment, strongest in France, that went with atheism, materialism, hedonism, and with secret societies of Illuminati, Freemasons and Jesuits – and, after the order was dissolved for a time by the pope in 1773, ex-Jesuits, equally sinister. Whereas the poet Alexander Pope had written that what-ever is, is right, it seemed to radicals that whatever is, is wrong. As material-ists and devotees of scientism (or even as straightforward Cartesians), they might also see animals as mechanisms, to be experimented on and vivi-sected with what strikes us as casual brutality in often rather inconsequen-tial physiological studies. Even worse, that notion might be extended to humans: the Parisian surgeon Julian Offray de La Mettrie (1709–51) famously published *L'Homme machine* in 1747, as a result of which he had to decamp to the court of Frederick the Great in Berlin. It is true that we are not unlike machines, becoming aware as we get older of our creaky mechanisms; but humans, dead or alive, are not machines, though they were and may still be made subjects for experimental medicine, and perceived as mere hands in sweatshops and dark Satanic mills. The serious threat that people could be seen and treated as objects lay behind Immanuel Kant's (1724–1804) ethical principle that they must always be ends, never means.

Many of those involved in the Enlightenment never subscribed to radical and reductive world-views, but were moderates, Deists for whom Freemasonry was a charitable and convivial activity enlivened by harmless little rituals. Moderates perceived order and design rather than blind watchmaking throughout nature; for them, living and dead matter obeyed different laws (vitalism), and minds and bodies were distinct. In politics they were Whigs, seeking piecemeal improvement, tinkering, effecting checks and balances, extirpating corruption but not turning the world upside down. In our own day, though there have been radicals among them, most scientists most of the time have supported the status quo, and the same was probably true in the past: respect for nature and nature's laws may make one law-abiding. During the Enlightenment, academies and the Royal Society represented part of what is nowadays called the Establishment; and established, serious-minded natural philosophers and medics supported it while favouring reforms. Meanwhile, outsiders turned radical: in England, Priestley and other Dissenters; and especially in France, journalists, popularisers, lawyers and those rejected by the Académies, who had grievances against their aristocratic, arbitrary and absolute government, *philosophes* whom we might call 'public intellectuals'.

All involved in the Enlightenment hoped to bring into being a modern world of limitless progress as a scientific mindset came to prevail more and more widely. Common sense was being increasingly refined, trained and organised, and even inspired. After a couple of centuries the Scientific Revolution was going from strength to strength, unlike those previous bursts of interest in the natural world and its workings, in Mesopotamia, ancient Athens and Alexandria, China, Islam and the Yucatán, which had not been sustained. There were five good reasons for its acquired impetus: a flow of able young recruits, and government support; its association with international trade and communication; its promise of bettering the human condition, and of real and endlessly growing understanding where religious dogma and ancient authority had failed; its ever-closer links to booming new industries; and, most importantly, its enduring institutions. Major upswings and downturns, spasmodic interest followed by sclerosis, seemed things of the past. Universities that had primarily been places where clergy, lawyers and doctors were prepared for their professions (and young grandees polished) began, first in Prussia from 1727, to train students formally in 'cameralistics', a course including science and its applications, ready for careers as civil servants. At Oxford and Cambridge, voluntary extracurricular lectures on chemistry and its usefulness were crowded. In France, technical schools were founded to train engineers, and in Saxony a mining academy was established.

Teaching meant textbooks; societies in provincial towns and cities held lectures, built up a library that might well be the nucleus of their activity, and aspired to publish journals like those that propelled metropolitan science; in London, the Royal Society, and in Paris, Berlin, Stockholm and St Petersburg, the academies, supported the publishing of scientific books; and enterprising publishers brought out both popular science for wide readerships and handsome natural histories for the well-off. Printing had been crucial in maintaining the impetus of the Scientific Revolution, as of the Reformation and Counter-Reformation; during the eighteenth century printing, publishing and bookselling began to separate from each other, as the world became more specialised. The market for books was still relatively small and risky, and books remained expensive, especially as they were issued in paper wrappers and the buyer was expected to have them bound in accordance with his or her personal taste. One way out was through subscription publishing, where the author found enough people prepared to commit themselves to buying the book in advance of publication; a list of the subscribers' names would be printed at the front of the volume. Thus in Borlase's *Natural History of Cornwall* over three hundred subscribers are listed: amid local gentry and serious naturalists in Britain and abroad, we find the aesthete Horace Walpole and Samuel Johnson's friend Henry Thrale (c. 1728–81), a wealthy brewer. A refinement of this system was to issue sumptuous works of natural history in parts, over months or years, making them seem less expensive: the receipts from part one paid both royalties and the printing costs of part two, and so on, and when the work was complete the purchaser would arrange the parts as advised or in accordance with his or her taste and have them bound.

Although printing techniques changed little, and Benjamin Franklin's training would have been very like Plantin's, texts printed in the eighteenth century look to modern eyes much more familiar and elegant than earlier ones. Outside Germany, blackletter fonts modelled upon scribal conventions had disappeared. As with architecture, the history of typefaces is a matter of innovation and revival: we have become increasingly eclectic.

Two that have made a comeback are our ubiquitous Times New Roman, modelled on a sixteenth-century font from the Plantin-Moretus foundry, and the clear roman font designed by Claude Garamond (1490–1561) in Paris that was a favourite for two hundred years. Another font that printing historians describe as 'Old Face' is the graceful design by William Caslon (1692–1766). The lighter, less robust 'transitional' typeface of his younger contemporary John Baskerville (1706–75) of Birmingham, with finer vertical lines adapted to smoother 'wove' paper, was much admired in its time and since; while 'modern' fonts with increasing contrast of thick and thin strokes made practicable by further improved technique and paper quality began in France and Italy with the family firm founded by François Didot (1689–1757), and the elegant fonts of Giambattista Bodoni (1740–1813). All of these are easy on the modern eye, do not draw particular attention to themselves and are comfortably readable in texts of book length. Catchwords at the bottom of the page, the letters i and j and u and v not being distinguished, and the long s, customary until about 1800, confuse some twenty-first-century students (and also computers). The ligatured double s survives in German typefaces, and in handwriting the long s (very like an f) persisted longer than in print. The marginal notes characteristic of earlier texts disappeared in the eighteenth century; footnotes elaborating on the text, and (patchily) references and bibliography, gradually became customary in learned works as it became accepted that one should pay one's intellectual debts in full and in public.

On Bacon's list, printing was one of the inventions that he thought would prevent further 'dark ages' of ignorance from following upon barbarian conquests. The others were the magnetic compass, making ocean voyages and trade practicable, and gunpowder, which would not only repel barbarian hordes but also be of great value in engineering and mining. By the end of our period, the compass, in conjunction with other navigational aids like the sextant and the chronometer, had indeed opened up the seas to trade and empire, enriching Europe and propelling technical change in workshops, shipyards and arsenals. And muzzle-loaded guns hauled by horses or mounted on ships, and requiring new levels of skill in casting and boring metal, were a feature of the warfare that went with European rivalries at home, in America, in India and in the Balkans, where the

Ottoman Empire was beginning to give ground. European armies clad in increasingly splendid and intimidating uniforms acquired muzzle-loading muskets to fire volleys of lead balls towards each other at a range of about 160 feet (50 m), following up with bayonets. Thus flintlock Brown Bess or Tower muskets were used by the British army for over a hundred years from 1722, the Tower of London being the armoury where examples were kept to ensure standardisation. By the latter part of the seventeenth century, guns with rifled barrels were being made, giving the ball a spin and a more accurate trajectory. For hunters, rifles with backsight and foresight for taking careful aim were well worth having. In warfare, they were issued to sharpshooters to pick off targets; in America, the 'minutemen', rebellious colonists ready to drop everything at short notice and take pot shots at redcoats with their rifles, became heroes. In naval battles, marksmen in the rigging would aim at officers on the quarterdecks of hostile ships, most notably Horatio Nelson (1758–1805) at Trafalgar (1805). Being up to date in technology became more than ever necessary in the competition between states for wealth and power; and science was inextricably linked to this process.

Science was also a vital part of intellectual culture. Like other parts, it had its funny side. Pepys reported of Charles II's opinion of the Royal Society: 'Gresham College he mightily laughed at for spending time only in weighing of ayre, and doing nothing else since they sat.'[1] With hindsight, we know that experiments on air by Boyle and Hooke were not only striking, but worth the attention they got; but a gaggle of dilettantes and absent-minded professors are always worth a giggle. Though widely respected, doctors were good for a laugh too; Molière in his comedies *Le Médecin malgré lui* (1666) and *Le Malade imaginaire* (1673) mocked the learned language and pseudo-explanations of physicians, like the 'virtus dormativa' characteristic of opium. In his *Le Bourgeois Gentilhomme* (1671) the central character, M. Jourdain, employs an absurd philosopher as tutor: offered physics, meaning the causes of meteors, rainbows, will-o'-the-wisps, comets, lightning, thunder, thunderbolts, rain, hail, wind and whirlwinds, he remarks that there's too much commotion and hullabaloo there. Instead he learns phonetics, and discovers that he has been speaking prose all his life.

In London, where Ben Jonson had got laughs from alchemy, his admirer Thomas Shadwell (1642–92), poet, playwright and Whig who became poet laureate in 1689, got them too with his comedy *The Virtuoso* (1676). Sir Nicholas Gimcrack (whose name gives away his character) is an enthusiast for bottling air and for transfusing blood from sheep to men; he is ruined by the expense of his experiments, and bewildered by the life around him, but remains in hope of finding the philosopher's stone. In the event, the play's pastoral charm has outlasted its carefully aimed satire, said to have humiliated Hooke: what Shadwell is remembered for now is his song 'Nymphs and Shepherds'. Another squib aimed at the Royal Society is in *Gulliver's Travels* (1726) by Jonathan Swift, where the third extraordinary kingdom encountered, the flying island of Laputa, is inhabited by dotty inventors bottling sunshine extracted from cucumbers, and taking measurements for clothes by sextant rather than tape measure. Gulliver's other adventures, satirising political partisans, are better remembered, but Swift also knew what men of science were doing and sent them up. Men of science were robust enough to stand up to a bit of mockery, and were not put down. There was nothing like the twentieth century's 'two cultures': circles overlapped, and science was one activity, hobby or interest among others, which very few would hope, expect or want to make full time, or their profession.

Assiduous men of science, no longer feared as necromancers, might indeed be suitable targets for wit; but science was a serious subject employing tested theory and explanation, a coming to grips with the wonder of the world, a topic of intelligent conversation in salons and coffee houses, a part of culture creating ripples beyond the circles of its devotees. As such, it was propagated by lecturers, professorial like Willem 'sGravesande (1688–1742) of Leiden, or itinerant like Desaguliers, his Huguenot translator. They both wrote up their lectures into books; others propagated science in the form of natural theology and in writings targeting women. Prominent in this genre was Fontenelle's book of 1686 on plurality of worlds, in which a philosopher expounds the new astronomy in dialogue with a marquise as they stroll in her garden in the cool of the evening. It was a huge publishing success on both sides of the Channel, with four English versions between 1687 and 1715, among them translations by Glanvill and by Aphra Behn (1640–89). The latter's career as a writer – more or less unprecedented for a woman – was colourful, though probably less so than her own account of it made it seem; in the 1650s she was perhaps a Royalist spy, in 1663 she spent some time in the new British colony of Surinam (captured by the Dutch in 1667), and back in Europe married a German merchant. Soon widowed, living by her pen, she wrote plays, poetry and prose, and was associated with the circle of the libertine Lord Rochester (1647–80). As a staunch Tory, supporting the claims of King James II, she was attacked by Shadwell; her unedifying career and associations also brought her into disrepute among the respectable. In the early 1680s she seems to have visited Paris, and Fontenelle's was among various works she translated from French. Because of Fontenelle's embrace of Venusians, Martians and other planetary inhabitants, his book was an object of suspicion among the religiously orthodox, though this did not diminish its popularity. The sixth English edition, of 1737, based on Behn's and three other translations, advised caution in theorising (those who have seen a galumphing elephant may disagree about the metaphor):

> In case of new Discoveries, we should not be too importunate in our Reasonings, tho' we are always fond enough to do it; and your *true Philosophers* are like *Elephants*, who as they go, never put their second Foot to the Ground, 'till their first be well fixed. The Comparison seems the more rational to me, *says she*, as the Merit of these two Species, *Elephants* and *Philosophers*, does not at all consist in exterior Agreements.[2]

To this edition was appended a translation of the oration in defence of the New Philosophy that Joseph Addison (1672–1719), essayist, playwright and politician, delivered in Oxford in 1693, beginning:

> How long, Gentlemen of the University, shall we slavishly tread in the Steps of the Ancients, and be afraid of being wiser than our Ancestors? How long shall we religiously worship the triflings of Antiquity as some do Old Wives Stories? It is indeed shameful, when we survey the *great Ornament* of the present age [Newton], to transfer our Applauses to the Ancients, and to take pains to search into Ages past for Persons fit for Panegyrick.[3]

Joseph Addison (1672–1719), in the *Spectator*, the magazine which he edited with his friend Richard Steele (1672–1729) from 1711 and distributed around Britain by mail coach, published stylish, accessible and delightful essays, especially social comedy with a series of characters, male and female, promoting bourgeois sentiment, reasonable religion, a relaxed civil society and modernity – very different from the raffish world of Rochester and Aphra Behn, or a radical Enlightenment.

In Addison's ode 'The Spacious Firmament on High' the stars and planets reveal the work of an almighty hand; and later the eminent physician and poet Mark Akenside (1721–70) wrote a hymn to science, published in another periodical characteristic of its time, the *Gentleman's Magazine*, in 1739.

> Science! Thou fair effusive ray
> From the great source of mental day,
> Free, generous, and refined!
> Descend with all thy treasures fraught,
> Illumine each bewilder'd thought,
> And bless my labouring mind.
>
> But first with thy resistless light,
> Disperse those phantoms from my sight,
> Those mimic shades of thee:
> The scholiast's learning, sophist's cant,
> The visionary bigot's rant,
> The monk's philosophy.
> Give me to learn each secret cause;
> Let Number's, Figure's, Motion's laws
> Reveal'd before me stand;
> These to great Nature's scenes apply,
> And round the globe, and through the sky,
> Disclose her working hand.[4]

Science as intellectual hygiene was an acceptable idea in a world putting the old-fashioned behind it, where explanation meant efficient cause, where in law courts the demands of science made witchcraft incredible and where religion was rational.

Not everyone believed that religion could be rational. In Edinburgh, David Hume (1711–76) not only argued sceptically that 'cause' meant no more than constant conjunction without any necessary connection, but in his posthumously published *Dialogues concerning Natural Religion* (1779) raised deep questions about God's wisdom and benevolence, threatening even Deism. Philo, one of the characters in the dialogue, remarks that the whole Earth seems cursed and polluted – something with which evangelicals might have agreed. But he went on to add:

> Perpetual War is kindled amongst all living Creatures. Necessity, Hunger, Want stimulate the strong and courageous: Fear, Anxiety, Terror agitate the weak and infirm.... Weakness, Impotence, Distress attend each stage of that Life: and 'tis at last finish'd in Agony and Horror.... Curious Artifices of Nature ... embitter the Life of every living Being.

Warming to the theme of the dysteleology all too present in the natural world if we look behind the smiling face of nature on a spring day, he noted: 'The whole presents nothing but the Idea of a blind Nature, impregnated by a great vivifying Principle, and pouring forth from her lap, without Discernment or parental Care, her maim'd and abortive Children.'[5] He was seeing what Tennyson would later call 'Nature, red in tooth and claw'. Other characters in his dialogue make the case for Intelligent Design, but Hume shocked contemporaries by dying peacefully without the consolations of religion. His philosophical writings roused Kant from what he called his dogmatic slumber and set him thinking about the limits of reason; but in Britain it seemed paradoxical to deny that causation was a real and necessary connection, and he was celebrated for his *History of England* (1754–62). It was only in the nineteenth century that Hume was appreciated as a philosopher, especially by Huxley, who admired his mitigated scepticism, seeing it as in line with scientific method, and as the basis of his own position – for which he coined the word agnostic.

In Britain, Hume had had few overt followers; after all, there was the option of Unitarianism, a 'feather-bed to catch a falling Christian', as it was described in Darwin's family. In France it was different. Voltaire, in exile in England in 1726–9, had been impressed by British constitutional government, by the literary, intellectual and political circles in which he moved, and by Newton's funeral in Westminster Abbey. Back home, in bad odour again with the powers that be, in 1734 he went to live with a marquise, Gabrielle Emilie du Châtelet (1706–49) and her complaisant husband: there, in 1738, he wrote a brilliant popular account of Newtonian philosophy, dedicating it to her. As a girl she had met Fontenelle and talked astronomy with him; later she was tutored by Maupertuis, worked with 'sGravesande and published work on physics. When she began her relationship with Voltaire, she undertook the difficult job of translating Newton's *Principia* into French, with a commentary: it was published after her premature death in childbirth. Voltaire's more accessible book was at once translated into English, making 'these Elements easy and intelligible to those, who know no more of *Newton* and Philosophy than their name'. Conducting his readers through the realms of light and gravity, he mocked his countrymen's love of little 'systems', with Descartes' in mind, echoing Newton's empirical demand for *verae causae* in writing of 'attraction':

> We are now to enquire, what are the Effects of this Power; for if we can make any Discovery of its Effects, the Existence of it will be evident. Let us not begin with assigning Causes, and forming to ourselves Hypotheses; that is the sure Way to wander from the Mark: Let us pursue, Step by Step, that which really passes in the System of Nature. We are Voyagers, arrived at the Mouth of a River, and must labour up the Stream, before we guess at the Situation of the Source.[6]

The Dutch were already Newtonian; Maupertuis, La Condamine, Voltaire and Emilie du Châtelet turned the French into Newtonians also, carrying his work and that of Leibniz forward as the powerful new tool of mathematical analysis, and making him into an august and representative figure very different from the rather crabbed original.

In Italy women had been present if not prominent in the proliferating and various academies of the sixteenth and seventeenth centuries; and when Descartes' patron Queen Kristina of Sweden abdicated in 1654 and went to

Rome she became an important patron there also. As science became something everyone should know about, more women were drawn in, and books were written aimed specifically at them. In 1733 Francesco Algarotti (1712–64), a young Venetian polymath who had studied in Bologna and Florence, visited Paris, where he met Voltaire and Emilie du Châtelet, charmed his hosts and became enthused with Newtonian philosophy. In London he was elected FRS in 1736; he spent some years in Germany from 1739 with Frederick the Great, and was entrusted by Augustus of Saxony with forming his art collection in Dresden, for which he bought paintings and displayed them according to historical principles. Well known also as an essayist and music critic, Algarotti had a wide circle of correspondents. In 1737 he published a dialogue on Newtonian philosophy, especially light and colours, aimed at women: the English translation of 1739 was called *Philosophy Explained for the Use of the Ladies*. The translation was anonymous, but it was, very appropriately, by a woman, Elizabeth Carter (1717–1806). A clergyman's daughter, she was very learned, formidable in her scholarship and linguistic skills, strong-minded and independent, refusing offers of marriage, holding her own amid a wide circle of powerful and intellectual men. Though she rated her classical translations higher, with her version of Algarotti she played an important part in opening science to women, and was a pioneer 'bluestocking', making writing a respectable career for them in what remained a man's world. Women did not in France and Britain have the chance to enter academies or the Royal Society, but some women (especially widows) carried on skilled trades and businesses, and others assisted their husbands in the library, sorting collections, in the observatory or in the laboratory; some drew, etched or coloured illustrations in scientific publications. Among the gentry, some ladies hosted salons, while others translated texts, for girls might learn modern languages while their brothers struggled with Latin and Greek; some assisted in research and publication. But it was not easy to break in to this world, and science has always favoured a macho rhetoric of warfare against ignorance, superstition and disease, open to modern feminist critiques as patriarchal, exploitative, abstracted and reductively mechanical.

Leibniz had begun his controversy with Newton and his mouthpiece Samuel Clarke (1675–1729) in a letter to Princess Caroline of Anspach

(1683–1737), leaving Hanover for London with her husband (the future George II (1683–1760)) when her father-in-law succeeded to the throne as George I (1660–1727); an intellectual, she established a salon at Leicester House. The resulting exchange of letters, published in 1717 after Leibniz's death, was an important landmark in philosophy of science, with two outstanding mathematicians contesting ideas on space and time. Leonhard Euler, born in Switzerland and working in Berlin and St Petersburg, was perhaps their equal in the next generation, and he corresponded (uncontroversially) with the Princess Friederike of Anhalt-Dessau (1745–1808) about mathematics. These *Letters to a German Princess*, published in French (1768–72) and soon translated, became a classic exposition of mathematics and natural philosophy by an expert for a lay readership. Women of sufficient social status could thus not only be patrons in the scientific enterprise of the eighteenth century, but also more seriously be participants like Emilie du Châtelet, or catalysts like the princesses Caroline and Friederike.

Not everyone was serious about their life or investment in that century, which featured, along with boozing and gambling, booms, credit 'bubbles' and optimistic 'projectors' seeking funding, spin-offs from science. But behind these was a real economy involving new and expanding industries, and one of its first needs was easy transport. The Romans built good roads for moving their forces around, and for armies with ordnance the need was even more urgent. To prevent Bonnie Prince Charlie's (1720–88) Highlanders invading England in 1745, the elderly General George Wade (1673–1748) was stationed in Newcastle, in the east; heading west to intercept the Scots, his guns became so hopelessly bogged down that he missed them both then and on their subsequent retreat. After the previous Jacobite rising in 1715, he had built a network of military roads including forty stone bridges in Scotland; as a result of this debacle, a road was built across England from Newcastle, much of it along Hadrian's Wall (recycling the stone as ballast). Bad roads were an obstacle not only to armies, but also to trade. Civil engineers built the roads that, leased to turnpike trusts charging tolls, greatly increased the speed and reliability of road transport, though the development proved tiresome for poor victims of science who had to pay to use roads that had formerly been free.

Specialist schools for engineers were established in Paris, and great technical projects undertaken by the state. Captured French warships were preferred by some British sailors as better designed because less rule of thumb had gone into their construction than was the case in Britain. War was as ever a great catalyst of science, knowledge being power, and throughout the eighteenth century France and Britain sporadically fought each other. Because most battles took place at sea or overseas, in North America, the West Indies and India, the wars did not devastate the combatants' home countries. Through them, and through the colonial conquests and trade that followed them, science in effect had its victims: aboriginal inhabitants were subject to cultural dislocation, displacement, violence, enslavement, taxation and disease, while among European soldiers and sailors in distant environments death rates were appalling. To-and-fro traffic in important commodities, however, continued: maize, potatoes, tobacco, Brazil wood for dyeing and Jesuit's bark from the Americas; wheat, sheep and horses (transforming Navajos, for example, into a nation of mounted shepherds) from Europe; coffee, indigo, tea and porcelain from Asia.

Ships and boats were much cheaper and more efficient than wagons for moving heavy loads, and networks of canals were dug all across Europe to extend and connect navigable rivers. In Britain they proliferated, becoming a vital factor in promoting industries after James Brindley dug the duke of Bridgewater's canal in 1761 to take coal to Manchester. To make locks, tunnels and aqueducts easier to construct in the hilly north of England, and because land was expensive, this and subsequent canals were made narrower than those elsewhere in Europe. The 'narrowboats' used on them, pulled by a single horse, could hold 30 tonnes and were thus much more economical than a wagon. They were also much smoother, and Wedgwood promoted and invested in canals to get his fragile wares to London and other cities, avoiding the breakages inevitable on jolting carts. Longer journeys with bigger cargoes could go, less smoothly, by sea. Cook learned his seamanship on board coasting vessels on the perilous North Sea, where the dangers of shipwreck were being slowly reduced by better charts, buoys marking channels and rocks, and the building of lighthouses.

What we call the Industrial Revolution was on most reckonings well under way in Britain, the first industrial nation, by the middle of the eighteenth century. But this revolution, like the Scientific Revolution, was a very long-drawn-out affair, depending on concatenations of circumstances, and by no means predetermined or obvious to all contemporaries. Though London was an enormous city by the standards of the day – the Great Wen (meaning a goitre or tumour), as it was known – eighteenth-century Britain was a rural country of small workshops. But just as communications were fast improving by the second half of the century, power greater than teams of oxen or horses was becoming available. Thus the coal dug from the duke of Bridgewater's estate and from ever-deeper pits in the north and midlands of England and in lowland Scotland depended upon the steam pumps that followed upon Newcomen's original design to keep the mines dry. Outside the mining districts, water power remained the main resource, and when, in 1771, Richard Arkwright (1732–92) set up the first textile mill, it was water-powered like numerous small-scale mills involved in grinding corn, fulling cloth and pumping. Especially in the Netherlands and East Anglia, wind-mills did these jobs, as they had for centuries, attaining several horsepower on breezy days; but both wind and water power depended upon the weather, and were thus unreliable for large-scale industry. Otherwise, apart from pumping with Newcomen engines, energy much greater than harnessed animals could produce still lay in the future, with James Watt and his successors; and horses pulled the barges, coaches, carriages and wagons along canals, on roads and on the first railways or wagonways.

From the early years of the century, worldwide trade, and improving transport nearer home, expedited by science and engineering, brought consumer goods to an increasing number of people. A leisured class began to emerge, with time to spare for polite learning and rational recreation, and money to buy books, china, silverware or furniture, and to pay subscriptions to societies and libraries. Shopping became a pastime, and fashion important: fashion plates were printed, and dolls dressed in the latest style sent from Paris to the American colonies. People also had more time to be ill, and money to pay to be cured. Fashionable doctors could do very well, given a good bedside manner, and remedies such as mercury for syphilis, Jesuit's bark for malaria and opium for controlling pain began to do some good. Outside the realm of establishment medicine, patent pills, nostrums and quack treatments for sexual dysfunction flourished.

François Fénelon (1651–1715), a Catholic priest and celebrated preacher who had been tutor to Louis XIV's grandson, and archbishop of Cambrai since 1695, upset the king in 1699 when he published his satirical *Les Aventures de Télémaque*. Telemachus was the son of Odysseus, and the book is supposedly about his education at the hands of his tutor, Mentor, who dilates upon the immoral luxury of the court and its aggrandisement through endless wars of conquest. Fénelon was banished to his diocese, where he was beloved, but not effectively answered until Bernard Mandeville (1670–1733), a Leiden-trained physician, took up the cause of luxury and waste in his *Fable of the Bees* (1723). In verse, Mandeville described a hive in which all the bees, who had been living in a happy and prosperous state, resolved to live simply: demand fell, unemployment and dissatisfaction rose, and the result was disastrous. The would-be virtuous but now grumbling hive disintegrated, and the surviving bees, faced with recession and austerity, betook themselves to solitary life in hollow trees. The conclusion was that private vices confer public benefits. The book horrified contemporaries, who saw it as an attack upon Christian virtues; but nevertheless they carried on seeking more comfortable lives, promoted by science and technology – not only when shopping, but notably also when seeking to restore their health, particularly in spas and by the seaside.

Assayers had been testing the precious metals gold and silver for centuries, and after the wars of the early seventeenth century, when much silver had been melted down to pay soldiers, with rising prosperity there was a boom in silversmithing in France, and in Britain, where silver was hallmarked to confirm its purity. Those in France who groaned like Voltaire beneath the absolutist government looked with admiration at Britain, but France with its luxurious court was the cynosure of all eyes when it came to the arts of civilisation. The French Baroque and Rococo styles, and the splendid finish demanded by court circles, were greatly admired and aped in Britain. The Huguenot Paul de Lamerie (1688–1751), brought to England as a baby by his refugee parents, introduced French skills to British craftsmen as richly decorated silverware became a status symbol among the wealthy – indeed, Georgian silver has retained its desirability for collectors to this day. Like the clocks and watches made at the same time, it demonstrates the delicacy and virtuosity of those craftsmen. Lamerie and the top instrument-makers, like the painters of the Renaissance, employed assistants and outworkers under their supervision, meaning that many of their products were 'studio' rather than individual pieces. A similar division of labour happened in Birmingham, which became the great centre for the making of cheaper metal goods for the emerging middle class, spreading the taste for manufactures among newly affluent consumers. The hive remained happy. Science went (and goes) with wealth: while it promises prosperity, it also requires it – it is a form of long-term investment impossible in a hand-to-mouth world.

While Paris was the top place for formal science, London was the great centre for instrument-making, with British industry growing fast. Dissenters, tolerated second-class citizens excluded from Oxford and Cambridge and government, were especially prominent in promoting commerce, industry and modern education for both boys and girls. The Lunar Society of Birmingham is the most famous group of men concerned with these things, mostly Dissenters who met socially (at the full moon, so that they could see to get home) in the second half of the eighteenth century: the members included the physician and polymath Erasmus Darwin, Priestley, Boulton, Watt and Wedgwood. Very informal, in effect a dining club meeting in each other's houses, they discussed all sorts of topics, against a background of expanding business in china, in metals and in improved steam engines. Watt had taken out patents: the latter term had ceased to mean monopolies granted at the royal whim, but he nonetheless found that they were very difficult and expensive to defend in the courts against those who infringed them, and the patent system as we understand it was still in its infancy. The quality control that Wedgwood and Boulton aimed to achieve threatened to de-skill those who worked in their companies, whose judgement was replaced by instruments – making them too perhaps victims of science, along with midwives repressed by obstetricians and farm workers by improving landlords.

One epoch was giving way to another, a way of life being subsumed. Those who learned new skills could do very well: the middle class was expanding and the well educated were in demand as, with increasing prosperity coming from booming industry, experience disproved the old notion that teaching the poor was harmful and unnecessary, giving them expectations that could not possibly be fulfilled. But this was no longer a static world where ploughboys educated and sent to university might rob the sons of lawyers, parsons and doctors of the professional jobs which they were expected to fill. Nevertheless, social mobility turned out to go both ways after all, as conservatives feared. Protestants felt that reading was important so that everyone had access to the Bible, newly available in vernacular translations; but learning to read was by no means universal and writing was taught even less. So while there were poor boys who became eminent in science, it was much easier for those in comfortable circumstances. Similarly, apprenticeships for skilled trades, a good route into the sciences, normally required a premium to be paid by somebody.

The reach of science, the circle of knowledge, continued to expand as new phenomena were brought into its empire of classification, understanding, prediction and control. Newton had hoped to find laws in the realm of corpuscles akin to those he had demonstrated in the Solar System. That did not happen in the eighteenth century, but following the experimental route he had laid down in optics the realm of physics was expanded, though not quite as far as Molière's philosopher thought. Newton extended the mathematical method employed in optics since Euclid by judicious experiment. As with gravity, he showed how phenomena could be predicted by laws; although there was dispute about just what light might be, Newton opting for particles, Hooke and Huygens for pulses or waves. After his death, and following hints in the 'Queries' he appended to his *Opticks* in successive editions, others looked for new worlds to conquer.

It had long been known that amber or glass when rubbed will attract little pieces of paper, and perhaps even crackle or spark. This phenomenon, called electricity, seemed more of a parlour trick than a bit of serious science. It was attributed to an effluvium or vapour that would drive away or attract light objects: amber and glass, insulators, would hold but not transmit electricity, but metals would conduct it somewhere else. Like magnetism, electricity was a polar phenomenon: 'resinous' and 'vitreous' bodies were complementary and would spark when rubbed and brought close to each other. Before 1663 Guericke invented the first 'electrical machine', in which a globe of sulphur was rotated against his hand and became charged to a level that made more experiments possible. Newton suggested using a globe of glass instead of sulphur, and Hauksbee built machines with glass globes and woollen rubbers. In 1746 William Watson (1715–87) made one using a large glass disc or wheel. Such devices lent themselves to spectacular lecture demonstrations because, given insulation, a large electrical charge could be developed, producing massive sparks or making the operator's hair stand on end if he were insulated from the earth. If not, then such machines gave electric shocks, making muscles twitch involuntarily – not only in the living, but even in corpses, animal or human. Electrical machines joined air pumps as standbys for scientific lecturers with an audience to wow.

The globe or disc could be charged, but electricity could not be conveniently stored until, in 1745, the Dutch professor Pieter van Musschenbroek (1692–1761) invented the 'Leiden jar', a glass bottle filled with water and coated inside (and later outside as well) with metal foil. When this device was charged via a metal rod or wire in contact with the foil lining, it could be carried around; and when he touched the knob Musschenbroek got such a shock that he told his French friend Jean-Antoine Nollet (1700–70) that he wouldn't repeat it for the whole kingdom of France. Nollet boldly continued the research, finding very surprisingly that the thinner the glass, the more powerful the shock. The jar, or a 'battery' of them, could be used for entertainment: to amuse the king of France, a line of monks holding hands were given a shock that made them all jump simultaneously. Then, in 1747–9, Benjamin Franklin in Philadelphia, whose printing and publishing business had so flourished that he could contemplate retiring to pursue his two loves of science and public life, made Leiden jars and started to experiment with them. Working at the geographical fringe of the scientific world, reporting his work in letters to London, he made the curious discovery that the water was not necessary: a pane of glass, coated with foil on both its surfaces, would do just as well as the bottle. He suggested that the electric fluid was not just agitated by rubbing, but was accumulated in vitreous (positive) and diminished in resinous (negative) substances. When charged bodies touched each other or the earth, equilibrium was restored, perhaps with a flash or shock:

At the same time that the wire and top of the bottle, &c. is electrised *positively* or *plus*, the bottom of the bottle is electrised *negatively* or *minus*, in exact proportion: *i.e.* whatever quantity of electrical fire is thrown in at the top, an equal quantity goes out of the bottom. . . . The equilibrium cannot be restored in the bottle by *inward* communication or contact of the parts; but it must be done by a communication *without* [outside] the bottle . . . in which case it is restored with a violence and quickness inexpressible.[7]

This theory, coupled with Franklin's identification of lightning with electrical discharge and his recommendation of pointed metal lightning conductors, made his name as a man of science. He was, however, lucky to survive his experiment of flying a kite on a thundery day: atmospheric electricity was potent and soon had its martyr, Georg Wilhelm Richmann (1711–53), who was spectacularly electrocuted, creating a vacancy in the academy at St Petersburg.

Franklin's theory was developed and extended, and because electrical experiments were fun, many joined in and this one-fluid theory prevailed widely, especially in Britain. There were still problems. When a bullet goes through a sheet of paper it leaves a burr, a rough rim around the hole, on its exit; but an electric spark leaves burrs on both sides. Especially in France, the view gained ground that there must therefore be two electric fluids, positive and negative, flowing in opposite directions round the circuit. In 1767 Priestley, Dissenting minister, Christian materialist, political radical and friend of Franklin, published his *History and Present State of Electricity*, in which he sought both to bring some order into the field and to make his mark in the scientific world. In the preface he was prophetic:

Hitherto philosophy has been chiefly conversant about the more sensible properties of bodies; electricity, together with chymistry, and the doctrine of light and colours, seems to be giving us an inlet into their internal structure, on which all their sensible properties depend. . . . New worlds may open to our view, and the glory of the great Sir Isaac Newton himself, and all his contemporaries, be eclipsed by a new set of philosophers, in quite a new field of speculation.[8]

Getting into internal structures had been Newton's hope; now it became Priestley's programme. He wrote another big book on optics, but that did not go so well and he turned to chemistry, which had previously been a science of solid and liquid substances and their properties. Priestley would be the primary agent in changing that idea.

Fire, water, earth and air were the ancient elements, but now that science was getting below the surface to which mechanics was confined they were no longer fundamental: some fire was perceived as electrical (in lightning), some as chemical (in combustion) and some as vital (in animal heat). The economist and chemist Georg Ernst Stahl (1660–1734) of Halle saw fire as 'phlogiston' (from the Greek word for flammable), present in anything that would burn; his view caught on. Weightless, like light it was seen sometimes as a substance, and sometimes as what we might with hindsight call chemical energy – as was 'caloric', its successor in Lavoisier's new chemistry. Waters could be salt or fresh, more or less good for you, and some, like aqua vitae and aqua regia, were chemically distinct substances (brandy, and a potent mix of nitric and sulphuric acids). Farmers knew about rich and poor soils, but chemists were distinguishing 'earths' in terms of their constituents, finding new metals in the process. Everyone knew that the air of London and other cities, smoky, smelly and dirty, was less good than the clean air of the country or the seaside. Similarly, it seemed, air could be vitiated by breathing in a confined space: it became suffocating. Down wells, bad air accumulated; and in the famous Grotta del Cane on the slopes of Vesuvius at Naples humans felt all right, but low down there was a layer of bad air fatal to dogs unless they were speedily removed. Digging wells could be dangerous because bad air accumulated at the bottom, and Stephen Hales (1677–1761) advised throwing a shower of water down a well to disturb the air and improve its salubrity. One of the most prominent men of science in Britain, Hales had studied at Cambridge, where he remained for a decade as a Fellow of his college before becoming vicar of Teddington, not far upstream from London. He was an all-rounder: an exemplary cleric, he published sermons, became chaplain to the princess of Wales and a trustee and council member for the American colony of Georgia, and has a monument dedicated to him in Westminster Abbey. In Cambridge he was interested in John Ray's work, and with Stukeley made a machine to illustrate Newton's theory of planetary motion; in 1718 he was elected FRS. He devised a ventilator widely used in public buildings and on board ships, and received the Copley Medal in 1739 for his investigation of bladder stones. But it is for his book *Vegetable Staticks* (1727) that he is particularly remembered, 'staticks' meaning the dynamic equilibrium maintaining life. Distillation was an old practice involving the collection and condensation of volatile products, but Hales became interested in the air that came off when things were heated and that had been allowed to escape. He devised a means of collecting it by bubbling this air through water and filling an inverted glass. He found sometimes prodigious quantities, and was struck by how air was emitted and absorbed in chemical processes and in life:

> The air is very instrumental in the production and growth of animals and vegetables, both by invigorating their several juices while in an elastick state, and also by greatly contributing in a fix'd state to the union and firm connection of the several constituent parts of these bodies, *viz.* their water, salt, sulphur and earth.... our atmosphere is a *Chaos*, consisting not only of elastick, but also of unelastick air particles, which in great plenty float in it.[9]

Ordinary air was thus a mixture; parts of it were incorporated or fixed in various solid bodies, and might be released on roasting. Hales' work, copiously cross-referenced to hints in Newton's 'Queries' and tested in presentations to the Royal Society, was taken up by chemists, especially in Britain.

One of these was Joseph Black (1728–99), born in Bordeaux where his father was in the wine trade. He studied medicine at Glasgow and then Edinburgh, being much influenced towards chemistry by William Cullen; for his MD thesis in 1754, he took up the much debated question of why some alkalis are caustic while others are mild. Milk of magnesia (a mild one) was a remedy for indigestion, and Black showed that when it was roasted it gave off its fixed air and became caustic, like limestone turning to quicklime. When allowed to stand in the open air, the caustic magnesia (again like quicklime) gradually reverted to its mild form. He demonstrated quantitatively that the fixed air given off in the first stage of the cycle balanced the air absorbed in the second, and then that the fixed air was not the same as atmospheric air. It was this fixed air that had been observed in the Grotta del Cane or down wells, and it was a distinct substance, differing from respirable air in the same way that solids or liquids differ from each other. In 1756 Black moved back to Glasgow and became professor, and there his interest shifted towards heat. He befriended James Watt, then a university technician whose major interest was chemistry, backed his inventions and continued to correspond with him throughout his life. Black noted that large definite quantities of heat are required to melt a given quantity of ice at freezing point and again to turn water into steam at boiling point; and interpreted this as indicating that heat behaved like a weightless chemical substance. Water was a compound of ice and heat, and steam of water and heat. He found also that substances had a characteristic specific heat, required to raise a standard unit to a temperature one degree higher. In 1766 he returned to Edinburgh as professor, promoting the medical school's already high reputation, and spent much time in teaching large classes (many of those enrolled having no intention of practising medicine), in consultancy and in promoting industrial development. He was a major figure in the Scottish Enlightenment.

Priestley, brought up after his mother's death by his Calvinist aunt, attended Daventry Academy to train as a Dissenting minister, and from 1761 taught at Warrington Academy; for his work on electricity, urged on by Franklin and John Canton (1718–72), he was elected FRS in 1766. In the following year he became a minister in Leeds, and there visiting a brewery observed the bubbles rising to the surface of the fermenting beer, which, picking up from the work of Hales and Black, he identified as fixed air. A man of unbounded curiosity, great openness and enthusiasm, he took up the study of airs, claiming that when he began these experiments

> I was so far from having formed any hypothesis that led to the discoveries I made in pursuing them, that they would have appeared very improbable to me had I been told of them; and when the decisive facts did at length obtrude themselves upon my notice, it was very slowly, and with great hesitation, that I yielded to the evidence of my senses.... There are, I believe, very few maxims in philosophy that have laid firmer hold upon the mind, than that air (free from various foreign matters ...) is a *simple elementary substance,* indestructible and unalterable.[10]

At Leeds, and then from 1773 at Calne where he was employed as tutor and librarian by the prominent pro-American politician William Petty, Lord Shelburne (1737–1805), he isolated a series of distinct gases, 'factitious airs', over water (like Hales), and then, for those like ammonia that were soluble, over mercury; and invented soda water, the basis of the soft-drink industry. His most exciting discovery, in 1774, was the eminently respirable, vital or dephlogisticated air that we call oxygen. He saw science, truth that would set us free, as an important part of his armoury in promoting rational religion and democracy, freeing his country from priestcraft and kingcraft. His rhetoric of stumbling wide-eyed and innocent into new worlds need not be taken as gospel truth, but his wariness of hypotheses and delight in clear language were genuine, and he is one of chemistry's great discoverers and most genial characters.

Curiously, the undogmatic Priestley is remembered among chemists for refusing to abandon the theory that combustion involved the emission of phlogiston, rather than combination with oxygen as the great systematiser Antoine Lavoisier inferred in what he called his chemical revolution, in which he introduced the chemical nomenclature used ever since. Priestley saw little merit in hypotheses, believed that the new language was mere jargon and pointless neologism, and was content with the received theory. In fact, this much debated revolution was, like others, a process rather than an episode, and subsequent understanding has been distorted by the desire of participants and historians for a date and a hero. But stopping when we do in the current study, we can leave this science on the threshold of capturing the Romantic imagination with its dynamism, and the chance it offered 'by a closer observation of nature, to learn from what a small store of primitive materials, all that we behold and wonder at was created'.[11] One thing that made the 1770s memorable was that Priestley, Lavoisier and their colleagues and rivals used chemistry to account for physiological phenomena in some detail rather than in a very general way: in respiration, as in combustion, oxygen was absorbed and carbon dioxide emitted; and in photosynthesis plants in sunlight absorbed carbon dioxide and emitted oxygen. Chemistry and electricity seemed the key at last to understanding living organisms.

Science was proving itself to be a self-sustaining business, in which solving one problem disclosed others, and the advent of a theory of everything looked more and more of a chimera; the end was not nigh. As Priestley put it:

> The greater is the circle of light, the greater is the boundary of the darkness by which it is confined. But, notwithstanding this, the more light we get, the more thankful we ought to be. For by this means we have the greater range for satisfactory contemplation. In time the bounds of light will be still farther extended; and from the infinity of the divine nature, and the divine works, we may promise ourselves an endless progress of investigation of them: a prospect truly sublime and glorious. The works of the greatest and most successful [natural] philosophers are, on this account, open to our complaints of their being imperfect.[12]

Priestley reckoned that Newton or Hales would have had no idea what doubts, queries and hints his own work on air would have thrown up; and that he could not know how successors would build upon his discoveries. We know that about technology too: as it solves problems, it raises others. But we must be struck in this passage by Priestley's insistence on the sublime and glorious prospects for science: its very incompleteness engages our imagination, making it a source of wonder and liberation worthy of everyone's close attention and participation. Unlike Lavoisier, Priestley saw science as open to all, as his own experience showed: it was not something to be confined to a new priesthood of experts in expensively equipped laboratories. The austere Lavoisier bringing logic and quantities into chemistry, and the exuberant Priestley revealing the wisdom and benevolence of God in following experiment wherever it led, represent between them the main strands of the Scientific Revolution. Both were also involved in making science useful, Lavoisier through his gunpowder researches at the Paris Arsenal (where his laboratory was) and his model farm, and Priestley with his inventions (soda water, the india-rubber eraser and the timeline) and his close association with the Lunar Society of Birmingham. Lavoisier as a financier and tax-farmer, Priestley as a clergyman and tutor, were neither of them full-timers in science; while not amateurish, they cannot be called professionals. A new world was dawning that has been called the Second Scientific Revolution, in which professional scientists, trained in universities and technical institutions and perhaps taking higher degrees, became increasingly specialised, and industries arose based closely on up-to-date science. Priestley and Lavoisier stood at its threshold.

Davy, who had entered the Promised Land, looked back to his triumphant youth, 'when, full of power, I sought for power in others . . . when every voice seemed one of praise and love . . . and every spray or plant seemed either the poet's laurel or the civic oak – which appeared to offer themselves as wreaths to adorn my throbbing brow'.[13] Science had brought him admiration: but already there were those with doubts about the way the world was going. Priestley's rational religion and its associated science and politics appealed to intellectuals, but the emotional and soul-searching hymns of the melancholy poet William Cowper (1731–1800) and the 'enthusiasm' of Methodists and pietists meant a resurgence of feeling and conviction impatient of sceptical doubt and coolly rational natural theology. In Germany, Friedrich Schleiermacher (1768–1834), answering the 'cultured despisers' of religion, made it a matter of feeling and emotion rather than cool reason; and in England the liberal and ecumenical philosopher and poet Samuel Taylor Coleridge also emphasised the priority of faith:

> Belief is the seed, received into the will, of which the Understanding or Knowledge is the Flower, and the thing believed is the fruit.
>
> Assume the existence of God, – and then the harmony and fitness of the physical creation may be shown to correspond with and support such an assumption; but to set about *proving* the existence of God by such means is a mere circle, a delusion.[14]

In reaction against Enlightenment confidence, painters admired the sublime, artfully disordered English gardens making formal layouts look old-fashioned, writers sought folk tales, 'gothick' style took on for buildings and novels, and travellers looked for inspiration in untamed nature. In the Romantic period the word 'genius' ceased to refer to djinni and came to be applied to those who displayed a godlike creativity as poets, painters or men of science (like Davy). Priestley's and Davy's optimistic enthusiasm was not universally shared: many, including Davy's friends William Wordsworth and Walter Scott (1771–1832), expressed in their writings nostalgia (homesickness) for the world that was lost, and Mary Shelley (1797–1851) was provoked by Davy's rhetoric into writing *Frankenstein* (1818). The new world might seem hostile, inhuman and threatened by sorcerers' apprentices. The realm of Apollo, associated with the clear light of the Sun, went with logic and objectivity; that of Dionysius, god of wine, with inspiration and imagination. To Romantics, Apollonian science seemed restrictive and prosaic, turning the world grey under the pretence of enlightening it. Wordsworth wrote:

> One impulse from a vernal wood
> May teach you more of man,
> Of moral evil and of good,
> Than all the sages can.[15]

And John Keats (1795–1821) said that all charms fly at the mere touch of cold philosophy:

> Philosophy will clip an Angel's wings,
> Conquer all mysteries by rule and line,
> Empty the haunted air, and gnomèd mine –
> Unweave a rainbow.[16]

The rainbow had entered the dull catalogue of common things – thanks to Newton. In London a toast at an 'immortal' literary dinner in 1817 was 'Newton's health and confusion to mathematics'. This was not the way science had been perceived earlier in the eighteenth century. In an ode to Newton, James Thomson, author of the well-known poem *The Seasons*, wrote about rainbows and refraction:

> Did ever poet image aught so fair,
> Dreaming in whispering glades by the hoarse brook?
> Or prophet, to whose rapture heaven descends?
> Even now the setting sun and shifting clouds,
> Seen, Greenwich, from thy lovely heights, declare
> How just, how beauteous the refractive law.[17]

Wordsworth genuinely admired Newton's physics as well as Davy's chemistry, and Keats had trained as a surgeon; so their attitudes to science were complex, and it was reductive mechanical and static science that, in what was after all an Age of Wonder, they rejected in favour of a world of life and forces. Learning facts, the realm of 'understanding', was properly but a prelude to 'reason' and wisdom. While science might be mocked, especially when inventions failed to work or structures collapsed, there was no doubt that it could be a source of wonder, truth, beauty, harmony, transcendence and meaning, requiring imagination, giving sense and purpose to its practitioners. Then as earlier, science could never be a full-time avocation, and those engaged in it shared the social and intellectual life of everyone else, and were just as likely to write or enjoy poetry and delight in pictures. They did not think of themselves as scientists, a self-conscious and distinct group. Music had been a part of the quadrivium with other mathematical disciplines, and went easily with science; architects needed engineering skills; anatomy was vital for artists, and so was colour – art (fine as well as useful) and science were a continuum. But there was agreement that European culture, wealth and power had changed enormously in the previous three hundred years, and that zest for natural knowledge and its application lay behind that transformation. Everybody clearly lived in a very different world from that of Vasco da Gama, Copernicus, Paracelsus and John Dee, and we should conclude our story with some reflection on that difference, and on whether 'revolution' is an apt term to describe this long story.

CHAPTER 13

REVOLUTION, EVOLUTION
HOW, THEN, DID SCIENCE GROW?

IN 1492 COLUMBUS LANDED in a new world, and in 1776 European colonists in America successfully proclaimed themselves a new and independent state, a nation with a Manifest Destiny before them. Those years in which the USA and the colonies in Latin America were born and grew up were what Einstein called the 'happy childhood of science';[1] we have been calling them the Scientific Revolution. Priestley had seen himself as being on the threshold of a new era. Natural philosophy, he believed, had been the source of all the 'great inventions, by means of which mankind in general are able to subsist with more ease, and in greater numbers upon the face of the earth'. It involved a lot of trouble and expense, demanding the patronage of the wealthy to maintain its momentum, but, provided its devotees remembered that 'speculation is only of use as it leads to *practice*', power over nature through learning its laws, its progress was assured. He also believed in the morally uplifting value of science:

a [natural] PHILOSOPHER ought to be something greater and better than another man. The contemplation of the works of God should give a sublimity to his virtue, should expand his benevolence, extinguish every thing mean, base, and selfish in his nature, give a dignity to all his sentiments, and teach him to aspire to the moral perfections of the great author of all things.[2]

Our foray into the social history of science has indeed shown how collaborative it is, how dependent upon care and truth-telling – but also how intensely competitive. Newton refused to acknowledge Hooke's part in his theory of gravity, filched Flamsteed's data, and accused Leibniz of plagiarism over the differential and integral calculuses, setting up and packing a Royal Society committee to denounce his German contemporary: the three men detested Newton for his actions, and he in turn hated them. Science is a very human activity, where priority is all-important; ambition and suspicion are its dark side. But we must never forget that it can and should be a source of positive ethical values as well as of power and intellectual satisfaction. This is especially important as it has come, in some measure, to displace the religious traditions in which it was born and brought up.

In a much acclaimed lecture in London in 1802 Priestley's disciple Humphry Davy, then at the outset of a brilliant career in chemistry, proclaimed the dawn of a bright day in which science, applied to the common purposes of life, would bring prosperity to all and political revolution would be averted. The promises made by the pioneers of previous centuries were in his generation being fulfilled: he lived in exciting times. 'Science' at that time still meant, in English as in French, an organised body of knowledge, which could be in what we call the humanities. Scientific institutions were by the 1770s both commonplace and respected. First in revolutionary France, where the Académie Royale des Sciences (abolished as elitist, but soon reconstituted as the First Class of the Institut) was the world leader, then in Germany, and in Britain in the 1830s, those of Davy's and the next generation practising and applying these mathematical, experimental or descriptive disciplines (our physics, chemistry, biology, geology and engineering) began to feel themselves a group dedicating their lives to a common enterprise, even hoping to make careers in science. In 1832 the English word 'scientist' was coined to describe them.

Davy and his like expected, and it slowly happened, that the Royal Society would henceforth elect to its own ranks only those who had distinguished themselves professionally in science, becoming more like the Paris Académie; it would no longer be open to those merely interested in the discipline. In 1824 Davy became the first chairman, and Faraday the first secretary, of the Athenaeum, a club for intellectual gentlemen set up in a handsome classical building in London. The Athenaeum and other gentlemen's clubs, unlike the coffee houses of the Scientific Revolution, were stuffily closed to non-members. In this new world, as science became more refined, trained and organised, scientists themselves became patrons, and scientific institutions ceased to have the mix of classes and professions that they had previously had.

Science truly came of age in the nineteenth century, confident and well prepared for the cultural and economic dominance that was its destiny. Scientists had inherited from three centuries of childhood and adolescence a critical tradition and group feeling associated with institutions, publication, patronage and peer review; centres of excellence (Paris, Leiden, Uppsala, London, Edinburgh, St Petersburg); global ways of thinking; methods and examples of testing, classifying, investigating and explaining; instruments to extend the senses, and a devotion to precision in measurement; models of the exact fit of theory and observation in astronomy, mechanics and optics; a world-view based upon linear, efficient causation and indestructible corpuscles of matter; and confidence that knowledge meant power. The Scientific Revolution had brought them to maturity, but there were crucial features still lacking: the making of modern science was not a matter of painting by numbers the sketch that Galileo, Descartes, Newton, Linnaeus and the other founding fathers had prepared. Another revolution bringing in professional science would be needed to achieve that and, like other genuine revolutions, it took only one lifetime, in what historians call the Age of Revolutions. European linear time and rapid change enthralled Tennyson's hero in 'Locksley Hall' (1842):

Not in vain the distance beacons. Forward, forward let us range,
Let the great world spin for ever down the ringing grooves of change.

Thro' the shadow of the globe we sweep into the younger day;
Better fifty years of Europe than a cycle of Cathay.[3]

Progress was speeding up, and science and technology were driving it.

This Second Scientific Revolution happened while, from 1789, France went through Revolution, Terror, Directory, Consulate, Empire, Restoration, a hundred days of Empire again, Restoration again, Revolution again, and then, from 1830, greater stability under the constitutional 'July Monarchy' of Louis-Philippe (1773–1850). Amid all this ferment, Paris was where, right across the sciences, the leading exponents were to be found. At the natural history museum and its neighbouring zoo, plants and animals were classified in a natural system, in which comparative anatomy illuminated relationships between living and fossil species. Medical teaching was thrown open, shackles were removed from the mentally ill, systematic physiological experimentation began and Parisian surgeons, well grounded in anatomy, performed daring operations. Chemists extended Lavoisier's 'revolution' following his execution as a fat-cat tax-farmer in 1794, using his nomenclature to make sense of reactions; and the exact study of crystals was begun. Courageous balloonists established that the chemical composition of the atmosphere was constant, though it got dangerously thinner and colder as they drifted upwards. In physics, the methods of mathematical analysis instigated by Maupertuis and his contemporaries were extended; Newton's universe was demonstrated to be stable and to require no divine interventions; the flow of heat was reduced to mathematical equations (which turned out to be more widely applicable); and thermodynamics was born from the abstract study of steam engines. These things were taught in the newly founded Ecole Polytechnique, where admission was by competitive examination, students received grants and the professors were Academicians, so bringing research and teaching together in one institution: revolutionary changes indeed.

The French did not have it all their own way. In Germany, still fragmented into small states, recovered after the ruinous Thirty Years' War of the seventeenth century but reeling after defeat by Napoleon's armies, the new University of Berlin was founded in 1810 on the plan of Wilhelm von Humboldt (1767–1835), Alexander's brother, to forward both knowledge (*Wissenschaft*) and the development of individual students' talents (*Bildungen*). This became the model for the research university; Justus Liebig (1803–73) at Giessen invented the research school and the graduate student, vital to science ever since. The German chemical community had been brought together and built up through a journal, published independently of a society; and such publications, rapid, informal and often specialised, became a feature of science at this time. Indeed, specialisation was the order of the day. The old ideal of a liberal education, very similar to that of one's father and grandfather, was displaced by a vision of up-to-date knowledge of a delimited field. Many of the divisions between sciences made in this period, institutionalised in societies and threatening further fragmentation, remain with us today. Founded in 1831, the British Association for the Advancement of Science set boundaries that in the English-speaking world divided 'science' from arts, humanities and (later) social sciences. At this point, the need for someone who could follow the latest mathematics and explain the results of experiments to educated readers became apparent, and Mary Somerville emerged as the first in a new line of distinguished interpreters of science in the wake of Voltaire and Algarotti.

In Italy the connection between electricity and muscular motion was established, and in 1800 Alessandro Volta (1745–1827) published in the Royal Society's journal his discovery that dissimilar metals in water give rise to a steady electric current. As Davy put it, this sounded an alarm bell to the experimenters of Europe; soon Davy himself ascertained that chemical affinity was electrical, and used a battery of cells to decompose caustic potash, revealing the extraordinary metal potassium. That experiment entailed the use of terminals made from another wonder metal, platinum, and generated sparks and coruscations that flew around the room: chemistry was a spectacular science, attracting crowds to lectures. Meanwhile, evolutionary hypotheses had been floated in Britain and France, and the interconvertibility of electricity and magnetism was demonstrated. Science was no longer just promising but proving to be useful: coal gas was bringing illumination to streets and public buildings, making evening classes possible, while the gas we call chlorine was being used as a rapid means to bleach textiles that had formerly needed long exposure to the Sun – always problematic in Manchester. There, chemists with works laboratories were beginning to oversee the process of dyeing, improving the consistency, adhesion and fastness of natural dyes; pharmacy also was building ever closer links with chemistry, and drugs began to be advertised with claims about their purity rather than just with anecdotes about their curative value.

Isaiah had visions of a world in which 'the wolf also shall dwell with the lamb, and the leopard shall lie down with the kid', where 'every valley shall be exalted and every mountain and hill shall be made low, and the crooked shall be made straight, and the rough places plain'.[4] He saw a restored Jerusalem at its centre; and in the last book in the Bible, John saw 'that great city, the holy Jerusalem, descending out of heaven from God'.[5] Those of us who love mountains need not take Isaiah literally, but these visions of a transformed Earth, and the two contrasting cities, the earthly and heavenly Jerusalem, the city upon a hill, were exceedingly powerful to the imagination. That sort of renewal of the world would have to wait upon God's intervention, but men of science saw their own version coming. In 1802, a year of brief interlude in a twenty-four-year world war, Davy electrified his audience with a vision of such an imminent future:

> In this view, we do not look to distant ages, or amuse ourselves with brilliant, though delusive dreams concerning the infinite improveability of man, the annihilation of labour, disease, and even death. But we reason by analogy from simple facts. We consider only a state of human progression arising out of its present condition. We look for a time that we may reasonably expect, for a bright day of which we already behold the dawn.[6]

Science-based industry and commerce would bring the peace and prosperity foreseen by Bacon. Davy's own researches in tanning and farming, and then his safety lamp of 1815, invented in the laboratory for use down the pit, were steps in that direction.

The Scientific Revolution was erected upon the foundation of belief in God as the guarantor of law and order. The French revolutionaries overthrew the established Catholic Church, and although Maximilien Robespierre (1758–94) instituted the Festival of the Supreme Being, there were many avowed atheists thereafter in France, especially among scientists and their associated public intellectuals and journalists. Religion everywhere was becoming increasingly a matter of individual conviction rather than collective expression. For Auguste Comte, society and individuals were outgrowing religion and metaphysics and entering the era of positive knowledge, but the great generalisations that consolidated science came from Germany and Britain, where scientists were the children not only of the Enlightenment but also of the Romantic movement, which evoked wonder and intellectual boldness as well as nostalgia. Modern science would be built upon three pillars: atomic theory, conservation of energy (and the latter's decreasing availability) and evolution over deep time. There had been no more than inklings of these powerful unifying visions in the previous centuries, but they now became the basis of a scientific world-view within which religion was an optional extra – a matter of individual choice and preference, or even a vestigial organ like our appendix. The world had ceased to be sacred; now it seemed no longer manifestly a creation. Gassendi's corpuscles were undetectable, metaphysical; the atoms of John Dalton (1766–1844) in due course made detailed chemical explanations and predictions possible through understanding of the structures of matter.

Eighteenth-century puzzles about momentum and 'vis viva' conserved in mechanical processes, and the intuition of Davy and his contemporaries that electricity, chemical affinity, magnetism, heat, light and motion might all be expressions of underlying and indestructible force, were formalised into the principle of conservation of energy. The newly fundamental science we call physics was born as the study of energy and its transformations. To conservation was added the idea that energy was becoming less available, that the universe was steadily running down: time, which had been a feature of God's creation and would come to an end at Armageddon, now had a direction determined by science. Where, looking upwards, Copernicus, Digges and Pascal had been awed by the vast size of the universe, their successors in the nineteenth century gazed down into the dark abyss of deep time. Scientific archaeology, basing its methods upon geology, superseded antiquarian treasure-hunting. There was no longer any doubt that fossil bones and shells were the remains of creatures now extinct; and a theory of evolution explained the process by invoking only causes observable and still operating today. It also provided a happier note of progress to set against physicists' vision of decline into heat death as entropy and disorder inexorably increased.

By 1860 these crucial features of modern science were coming into place, so its very foundations were different from those of the science of the

1660s, when corpuscles, mechanics and a divine plan prevailed. But Priestley, in writing about the history and present state of electricity and optics, reinforced a pattern of learning and thinking about science through its development. To be sure who they were, scientists sought their ancestors. We select our past from the great flux of history, and like to see it culminating in us and our time. Thus Davy and his contemporaries looked back over three centuries, and placed themselves in a heroic and carefully selected tradition consisting primarily of navigators, chemists, physicians, surgeons, astronomers, natural historians, antiquarians, mechanics, surveyors, engineers, philosophers, parsons and improving landowners. These predecessors (not always highly regarded in their own time) could be seen to be the prophets of the polished ('polite') and newly industrial commercial society now emerging and prospering. Nobody could any longer deny Bacon's message that knowledge was power. Literary critics tell us that the end of a narrative gives shape to the whole: Bacon was compared to Moses, seeing from afar the Promised Land that by 1800 his successors were entering in strength.

We know that our society is industrial, dominated by technology. For most of us in the West, our ancestors' lives were unimaginably different from our own: subsistence farming eked out with hunting or poaching, interrupted by plague, famine and war. Science was an important factor in the move from kinship and grandee patronage towards meritocracy that is such a key element of modernity. Nowadays a dominant and inescapable feature of life, by the 1770s it was already clearly very important, whether people liked it or not. It was a drama too. It was an unending quest: Glanvill called one of his books *Plus Ultra*, 'Further Yet', and every fresh discovery or synthesis opened the door to more. The circle of light expanded and with it the circumference of darkness, and the frontier was a busy and exciting place to be. Unlike the American frontier, it would never be closed.

Newton indeed seemed to many to have had the last word about the planets (though we know better), and then and since some voyagers in strange seas have ended up in the doldrums, finding nothing really worthwhile left to do. Though he never saw Antarctica, Cook believed that he had gone as far anyone could go in the frozen north and south; the poles proved a magnetic attraction to the adventurous, and the continents which

Cook and his contemporaries had outlined on the map cried out for scientific exploration, and exploitation. There were failures: natural magic did not work, mineral remedies killed as well as cured, bridges fell down, there was no short cut to gold, artificial languages did not catch on, no solid evidence emerged for ghosts and spooks, the promises of astrology were not fulfilled. Of course, failures in science and technology can be particularly instructive, but it can also happen that babies get thrown out with the bathwater. The failure of astrology meant that interest in the constellations of fixed stars declined until William Herschel (1738–1822), later joined and assisted by his sister Caroline (1750–1848), came from Hanover to England with a German band, became an organist in Bath and built a reflecting telescope. First he discovered the new planet Uranus, securing royal patronage as Galileo had done by naming it 'Georgium Sidus', 'King George's Star', and going on to 'gauge the heavens' by plotting stars and nebulae.

Patronage was the mainspring of European society right through our period, for most of the time in the hands of the mighty. Charles II's patronage of the Royal Society brought no help financially, but it ensured prestige, respectability and independence in troubled times. At all levels, those with power or money expected to attract clients, whom they might oblige, protect, promote – or drop. But although in the Royal Society, as in Bath, rank was set aside, one ought still to know one's place. Patronage took different forms: a Fellow of the society would find himself in the company of bishops, of wealthy patients if he was a medical man, of ambassadors, government ministers, and grandees with great houses and libraries – all very useful to know for anyone with a career to make. In France, Academicians usually made up their number by election, so that there patronage was in the hands of scientists. That pattern spread as older forms of patronage came to be seen as 'Old Corruption' in England, and as universities in Scotland and in England freed themselves from church or local-government control. In Germany ministries in the various states were in charge, and rocking the boat (as in the Revolutions of 1848) might result in a professor's dismissal.

By the 1770s different visions of science were available: Priestley saw it as trained and organised common sense, open to all, making us prosperous and free; while for Lavoisier it was an elite activity to be carried on by the talented and paid for by everyone else. Parts of natural history such as studying the distribution of plants and animals, and of astronomy such as looking for comets, remain open to amateurs, whose contribution is essential. We still need citizen-scientists. Others need to know about it. Recondite science required (and requires) interpretation, and Fontenelle, Algarotti and Voltaire were vital to this process. Popularisation was serious and important in developing both a scientific world-view and an educated public, and it was not until much later that in scientific circles it snootily came to be looked upon as entertainment, froth upon the surface of arcane knowledge only intelligible through mathematics.

Science, and with it prevailing world-views, was transformed in the trauma of those revolutionary years, dominated by war and social change; but the new science was built upon the foundations we saw established in those earlier centuries. What then of revolutions? Mythologists note that satisfying explanatory stories are founded on social experience, and have applied such insights to the powerful myths, the grand narratives, that drive popular science. After 1917, those on the Left looked indulgently on the Soviet Union as the harbinger of the future. By 1962, when Thomas Kuhn (1922–96) published his book on scientific revolutions (the plural being significant), the rising generation had been children in the Second World War, and (while sceptical about the USSR) had seen in the 1940s and '50s revolutions against colonial powers and home-grown tyrannies, beginning in India and in Palestine, spreading to Egypt and Kenya, and culminating glamorously in Cuba and alarmingly in Mao's China. Patterns of enforced stasis and revolutionary change, from which Britain was one of the few places spared since 1688, seemed normal.

We are older and maybe wiser, and have seen how continuities are preserved even through the violent convulsions of political revolutions and, since 1989, how revolutions social and political that seemed final in their effects, like Communist rule, can be reversed. Perhaps, then, while continuing to speak of the Scientific Revolution for convenience, we should be ready to try a range of fresh metaphors, in the opportunistic manner of scientists explaining phenomena – maybe something more organic: growth or evolution. I tried twenty years ago to write the history of chemistry as a biography, in a manner one (complimentary) reviewer called postmodern; there one could see it rise from a childhood as 'chymistry' to a fundamental discipline, and then decline into the status of universal service science – though most of us hope that, unlike people, sciences will not die or fall to pieces.

Anyone who has written a biography, or indeed reflected upon his or her own life, knows how important the early, formative years are – though we need not believe that they are the happiest. The time we call the Scientific Revolution would then not, *pace* Einstein, be the childhood so much as the adolescence of science as it grew up and came to claim its important place in culture, before bursting into maturity in the nineteenth century. No model will fit the facts in the chancy, many-stranded story we have been looking at: revolution is too sudden and political, Fortune's Wheel too mechanical, growth too teleological, searching for truth too naïve, mythmaking too fuzzy, social construction too reductive, Marxism too schematic, opportunistic muddling along too vague. But all these have some part in our story, like the resonating structures in Linus Pauling's (1901–94) chemistry that yielded a more stable compound than any of them managed singly. But such infinite complexity will not do. We need something to help us sort out from history's booming, buzzing confusion a story worth following that illuminates the past and the present, and maybe 'Scientific Revolution', a label sanctioned by long usage, is as good as any thread to hold on to through the labyrinth of three hundred turbulent years in which the most powerful movement in the modern world grew up and evolved.

Notes

Chapter 1   Voyaging in Strange Seas

1. Elizabeth Knowles, ed., *Oxford Dictionary of Quotations*, 7th edn, Oxford: Oxford University Press, 2009, p. 574.
2. William Wordsworth, *The Prelude*, London: Edward Moxon, 1850, Bk 3, ll. 61–4.
3. Samuel Taylor Coleridge, 'The Ancient Mariner', II, stanza 5, in W. Wordsworth and S.T. Coleridge, *Lyrical Ballads* (1798), London: Penguin, 1999, p. 5.
4. Humphry Davy, *Consolations in Travel: or The Last Days of a Philosopher*, London: Murray, 1830, p. 35.
5. Steven Shapin, *The Scientific Revolution*, Chicago: Chicago University Press, 1998, p. 1.
6. Albert Einstein, 'Foreword', in Isaac Newton, *Opticks* (1730), New York: Dover, 1952, p. lix.
7. Genesis 2:5–3:24.
8. John Milton, *Paradise Lost* (1667), Cambridge: Cambridge University Press, 1976, Bk XII, ll. 646–9.

Chapter 2   The Deep Roots of Modern Science

1. Luis de Camões, *The Lusiads*, ed. G. Bullough, tr. R. Fanshawe (1655), London: Centaur Press, 1963, canto X, l. 74, p. 316.
2. C.R. Boxer, ed., *The Tragic History of the Sea, 1589–1622*, London: Hakluyt Society, 1959, p. 108.
3. Antonio Pigafetta, *Magellan's Voyage*, ed. and tr. R.A. Skelton, New Haven and London: Yale University Press, 1969, vol. I, pp. 54–5. W. Shakespeare, *The Tempest*, I, ii, 373.
4. Helen Wallis, 'The Patagonian Giants', in R.E. Gallagher, ed., *Byron's Journal of his Circumnavigation, 1764–1766*, London: Hakluyt Society, 1964, pp. 185–96.
5. Thomas Browne, *The Works*, ed. Geoffrey Keynes, London: Faber and Faber, 1964, vol. 2, p. 197.
6. Francis Bacon, *Sylva Sylvarum*, 7th edn, London: William Lee, 1658, p. 155.

Chapter 3   Refining Common Sense: The New Philosophy

1. Humphry Davy, *Consolations in Travel; or The Last Days of a Philosopher*, London: John Murray, 1830, pp. 234–5.
2. J.J. Thomson, *Recollections and Reflections*, London: Bell, 1936, p. 379.
3. William Lilly, *Christian Astrology*, London: John Partridge and Humph. Blunden, 1647.
4. Elias Ashmole, ed., *Theatrum Chemicum Britannicum* (1652), intr. A. Debus, New York: Johnson, 1967, pp. 342–3.
5. John Baptista Porta, *Natural Magick*, London: Thomas Young and Samuel Speed, 1658, p. 2.
6. Francis Bacon, *The Advancement of Learning and New Atlantis*, Oxford: Oxford University Press, 1960, p. 288.
7. Francis Bacon, *Sylva Sylvarum*, 7th edn, London: William Lee, 1658, pp. 211, 210, 155.
8. Thomas Sprat, *History of the Royal Society*, London: J. Martin, 1667, prefatory ode.
9. William Gilbert, *On the Magnet*, tr. S.P. Thompson, London: Chiswick Press, 1900, pp. 208–10.
10. René Descartes, *A Discourse of a Method for the Well Guiding of Reason, and the Discovery of Truth in the Sciences*, London: Thomas Newcombe, 1649, pp. iv, 14, 114.

Chapter 4   Looking Up to Heaven: Mathematics and Telescopes

1. Stillman Drake and C.D. O'Malley, ed. and trans., *The Controversy on the Comets of 1618*, Philadelphia: University of Pennsylvania Press, 1960, p. 310.
2. John Milton, *Paradise Regain'd* (1671), London: Vintage, 2008, Bk 4, ll. 40–2.
3. John Wilkins, *The Mathematical and Philosophical Works*, London: Vernon & Hood, 1802, vol. 1, p. 158; plurality of worlds, p. 13.
4. Henry More, *The Complete Poems*, ed. A.B. Grosart (1878), Hildesheim: Georg Olms, 1969, p. 93.
5. Bernard le B. de Fontenelle, *A Week's Conversation on the Plurality of Worlds*, tr. A. Behn, J. Glanvill, John Hughes, William Gardner, 6th edn, London: A. Bettesworth, 1738, p. 96.
6. C. Huygens, *The Celestial Worlds Discover'd; or, Conjectures concerning the Inhabitants, Plants and Productions of the Worlds in the Planets*, London: Timothy Childe, 1698, pp. 3, 13, 15, 150–1.
7. John Heath-Stubbs and Phillips Salman, eds, *Poems of Science*, London: Penguin, 1984, p. 138.
8. H.G. Alexander ed., *The Leibniz–Clarke Correspondence*, Manchester: Manchester University Press, 1956, pp. 90, 94.
9. F. Cajori, ed., *Sir Isaac Newton's Mathematical Principles of Natural Philosophy*, Berkeley: University of California Press, 1962, p. xxiv.

Chapter 5   Interrogating Nature: The Use of Experiment

1. I. Newton, *Opticks* (1730), intr. I.B. Cohen, New York: Dover, 1952, p. 376.
2. William H. Brock, *The Fontana History of Chemistry*, London: HarperCollins, 1992, p. 51.
3. J.R. Glauber, *The Works*, tr. Christopher Packe, London: Milbourn, 1689.
4. W. Charleton, *Physiologia Epicuro-Gassendo-Charletoniana*, London: Thomas Heath, 1654, p. 131.
5. R. Boyle, *Experiments and Considerations Touching Colours*, London: Herringman, 1664, Preface (p. iii).

6. R. Boyle, *New Experiments PHYSICO-MECHANICAL, Touching the Spring of Air*, 2nd edn, Oxford: Hall, for Robinson, 1662, p. 60.
7. D. Papin, *A New Digester*, London: Henry Bonwicke, 1681, Preface.
8. R. Hooke, *Micrographia*, London: Jo. Martin and J. Allestry, 1665, pp. 179–80.
9. Ibid., p. 210.
10. R. Hooke, *The Posthumous Works*, ed. R. Waller, London: Smith & Walford, 1704, pp. 24–6.
11. Ibid., p. 206.
12. Samuel Butler, 'The Elephant in the Moon', in John Heath-Stubbs and Phillips Salman, eds, *Poems of Science*, London: Penguin, 1984, pp. 95–6.
13. Clifford Dobell, *Antony van Leeuwenhoek and his 'Little Animals'* (1932), New York: Dover, 1960, p. 121.
14. Elizabeth Knowles, ed., *Oxford Dictionary of Quotations*, 7th edn, Oxford: Oxford University Press, 2009, p. 126.

Chapter 6   Through Nature to Nature's God: The Two Books

1. Michael Hunter and David Wootton, eds, *Atheism from the Reformation to the Enlightenment*, Oxford: Oxford University Press, 1992, pp. 40–1.
2. J. Boehme, *The Aurora*, trans. J. Sparrow (1665), ed. C.J. Barker and D.S. Hehner, London: Watkins & Clarke, 1960, p. 51.
3. T. Burnet, *The Sacred Theory of the Earth*, intr. B. Willey, London: Centaur, 1965, pp. 114–15.
4. H. More, *The Complete Poems*, ed. A.B. Grosart (1878), Hildesheim: Olms, 1969, p. 93.
5. J. Glanvill, *Saducismus Triumphatus: or, Full and Plain Evidence concerning Witches and Apparitions* (3rd edn, 1689), ed. C.O. Parsons, Gainesville, Florida: Scholars' Facsimiles & Reprints, 1969, p. 66.
6. J. Glanvill, *The Vanity of Dogmatizing*, London: Henry Eversden, 1661, p. x; see also pp. 240, 244.
7. R. Cudworth, *The True Intellectual System of the Universe*, London: Richard Royston, 1678, p. 888.
8. J. Ray, *The Wisdom of God Manifested in the Works of the Creation*, London: Smith, 1691, p. ix.
9. Ibid., p. 8.
10. N.A. Pluche, *Spectacle de la nature: or, Nature Display'd*, London: Pemberton, 1733, vol. 1, p. viii.
11. I.B. Cohen and R.E. Schofield, eds, *Isaac Newton's Papers and Letters on Natural Philosophy*, Cambridge: Cambridge University Press, 1958, pp. 280–1, 298, 302–3.
12. Acts of the Apostles 17:28.
13. [J. Toland], *Christianity Not Mysterious*, London: no publisher given, 1696, p. xx.
14. W. Wollaston, *The Religion of Nature Delineated*, 4th edn, London: Samuel Palmer, 1726, p. 59; Joseph Butler, *The Analogy of Religion*, new edn, London: Rivington, 1791.
15. Wollaston, *Religion of Nature*, p. 96.
16. Ibid., p. 215.
17. William Shakespeare, *Much Ado about Nothing*, V.i.35–6.
18. Psalms 14:1.
19. Edward Young, *Night Thoughts*, ed. George Gilfillan, Edinburgh: James Nichol, 1861, p. 262 (Night 9, ll. 772–3).
20. 1 Peter 5, 8, 9.
21. William Shakespeare, sonnet 73, *Complete Works*, ed. J.W. Craig, Oxford: Oxford University Press, 1954, p. 1116.

Chapter 7   Sharing the Vision: Scientific Societies

1. John Baptista Porta, *Natural Magick*, London: Young & Speed, 1658, preface.
2. *Essayes of Natural Experiments Made in the Academie del Cimento*, trans. Richard Waller (1684), intr. A.R. Hall, New York: Johnson, 1964, dedication.
3. John Ward, *The Lives of the Professors of Gresham College*, London: John Moore, 1740, p. ii.
4. Thomas Sprat, *The History of the Royal Society of London, for the Improving of Natural Knowledge*, London: J. Martyn, 1667, Cowley's ode, stanza 9.
5. Nehemiah Grew, *Musæum Regalis Societatis, or A Catalogue & Description of the Natural and Artificial Rarities Belonging to the Royal Society and Preserved at Gresham Colledge*, London: Rawlings, 1681, p. ix.
6. J.W. Goethe, *Italian Journey, 1786–1788*, London: Folio Society, 2010, p. 74 (5 October 1786).
7. Quoted in C. van Eck, *Classical Rhetoric and the Visual Arts in Early Modern Europe*, Cambridge: Cambridge University Press, 2007, pp. 186–7.
8. J. Herschel, *Preliminary Discourse on the Study of Natural Philosophy*, London: Longman, 1830, pp. 25–6; and *Treatise on Astronomy* (1833), new edn 1851, London: Longman, p. 5.

Chapter 8   Life Is Short, Science Long: The Healing Art

1. William Lilly, *Christian Astrology Modestly Treated in Three Books*, London: Brudenell, 1647, p. viii.
2. George Parker, *Mercurius Anglicanus; or The English Mercury*, London: MC, 1696; my copy.
3. *Medicina Antiqua*, ed. Peter Murray Jones, London: Harvey Miller, 1999, fol. 60v.
4. Genesis 2:9.
5. John Gerard, *The Herball or Generall Historie of Plantes*, ed. T. Johnson (1633), New York: Dover, 1975, unpaginated table at the very back of the book.
6. Charles Henry Hull, ed., *The Economic Writings of Sir William Petty* (1899), New York: Kelley, 1964, vol. 2, p. 346.
7. Laura Barber, ed., *Penguin's Poems for Life*, London: Penguin, 2008, p. 197.
8. Ambroise Paré, *The Apologie and Treatise*, ed. Geoffrey Keynes, New York: Dover, 1968, p. 132.
9. William Harvey, *The Circulation of the Blood* (1628), trans. T. Willis, London: Dent, 1932, p. 59; for a facsimile with French translation, *Etude anatomique*, ed. C. Laubry, Paris: Doin, 1950.
10. Ibid., p. 3.
11. Robert Burton, *The Anatomy of Melancholy* (1628), ed. F. Dell and P. Jordan-Smith, New York: Tudor, 1927, pp. 495, 576–7.

Chapter 9   Making Things Better: Practical Science

1. Neil Chambers, ed., *The Letters of Sir Joseph Banks: A Selection, 1768–1820*, London: Imperial College Press, 2000, p. 317.
2. Bernard Palissy, *The Admirable Discourses*, tr. A. la Rocque, Urbana, IL: University of Illinois Press, 1957, p. 192.
3. William Shakespeare, *Twelfth Night*, II.v.67.
4. Isaac Newton, *Mathematical Principles of Natural Philosophy*, tr. Andrew Motte, rev. Florian Cajori, Berkeley, CA: University of California Press, 1962, vol. 1, p. 6.
5. R. D'Acres, *The Art of Water-Drawing* (1659/60), ed. Rhys Jenkins, London: Newcomen Society, 1930, p. 9.

6. Joseph Moxon, *Mechanick Exercises on the Whole Art of Printing*, ed. H. Davis and H. Carter, 2nd edn, Oxford: Oxford University Press, 1962, p. xxv.
7. Joseph Moxon, *Mechanick Exercises*, ed. B. M. Forman, New York: Praeger, 1970, p. i.
8. Ibid., p. 119.
9. Moxon, *Mechanick Exercises on the Whole Art of Printing*, p. 11.

Chapter 10   The Ladder of Creation: The Rise of Natural History

1   William Turner, *Libellus de Re Herbaris, 1538, and The Names of Herbes, 1548*, ed. J. Britten, B.D. Jackson and W.T. Stearn, London: Ray Society, 1965, p. 146.
2   Edward Topsell, *The History of Four-Footed Beasts, Serpents, and Insects* (1658), intr. W. Ley, London: Frank Cass, 1967, conclusion of unpaginated Epistle Dedicatory.
3   John Lawson, *A New Voyage to Carolina* (1709), ed. H.T. Lefler, Chapel Hill, NC: University of North Carolina Press, 1967, pp. 131, 138, 156.
4   Thomas Browne, *The Works*, ed. G. Keynes, London: Faber, new edn 1964, vol. 2, pp. 30, 167.
5   Martinez Compañon, *La Obra, sobre Trujillo del Peru en el siglio XVIII*, Madrid: Ediciones Cultura Hispanica, 1978, vol. 6, plate LXXXIII.
6   Francis Willughby, *The Ornithology*, ed. and trans. J. Ray, London: Royal Society, 1678, unpaginated preface, pp. 4, 5.
7   Robert Plot, *The Natural History of Oxford-shire*, 2nd edn, Oxford: Oxford University Press, 1705, pp. 96, 112, 113.
8   Alexander Pope, *The Poems*, ed. J. Butt, London: Methuen, 1963, pp. 514 (I, 293) and 516 (II, 2).
9   William Borlase, *The Natural History of Cornwall* (1758), facsimile, ed. F.A. Turk and P.A.S. Pool, London: EW Books, 1970, p. v.
10 Olof Rudbeck, *Fågelbok, 1693–1710*, ed. B. Gollander, Stockholm: Eden Bokförlag, 1971, pp. 108, 32.
11 Edward Tyson, *Orang-Outang, sive Homo Sylvestris: or, The Anatomy of a Pygmie*, London: Bennet, 1699, preface, and p. 17.
12 William Shakespeare, *Henry IV, Part 2*, III.ii.46–9.

Chapter 11   A Global Perspective: Exploring and Measuring

1. William Falconer, *An Universal Dictionary of the Marine*, 4th edn (1780), Newton Abbott: David & Charles, 1970, p. 96.
2. Bernard le Bouvier de Fontenelle, *Elogium* (1728), in I. Bernard Cohen and Robert E. Schofield, eds, *Isaac Newton's Letters and Papers on Natural Philosophy*, Cambridge: Cambridge University Press, 1958, pp. 450, 454, 455.
3. Richard Walter and Benjamin Robins, *A Voyage round the World in the Years MDCCXL, I, II, III, IV, by George Anson* (1748), ed. Glyndwr Williams, Oxford: Oxford University Press, 1974, pp. 371–2.
4. Hugh Carrington, ed., *The Discovery of Tahiti: A Journal . . . Written by George Robertson*, 2nd series, XCVIII, Cambridge: Hakluyt Society, 1948, p. 229.
5. Louis de Bougainville, *A Voyage round the World . . . in the Years 1766, 1767, 1768 and 1769*, tr. John Reinhold Forster, London: Nourse, 1772, p. 299.
6. J.C. Beaglehole, ed., *The Endeavour Journal of Joseph Banks*, London; Angus & Robertson, 1962, vol. 1, p. 312.
7. J.C. Beaglehole, ed., *The Journals of Captain Cook, 1: The Voyage of the Endeavour, 1768–1771*, Cambridge: Hakluyt Society, 1968, p. 74.
8. J.C. Beaglehole, *The Journals of Captain James Cook, 2: The Voyage of the Resolution and Adventure, 1772–1775*, Cambridge: Hakluyt Society, 1969, p. 323.

9. J.C. Beaglehole, *The Journals of Captain James Cook, 3: The Voyage of the Resolution and Discovery, 1776–80*, Cambridge: Hakluyt Society, 1967, p. 696.
10. Engelbert Kaempfer, *The History of Japan*, tr. J.G. Scheuchzer (1727), Glasgow: MacLehose, 1906, vol. 1, p. xxxi.
11. Carl Linnaeus, *Lachesis Lapponica, or A Tour in Lapland*, ed. J.E. Smith, tr. C. Troilius (1811), New York: Arno, 1971, vol. 1, pp. 131–2.
12. Lewis de Bougainville, *A Voyage round the World*, tr. J.R. Forster, London: Nourse, 1772, pp. 174–6.
13. William Shakespeare, *King Lear*, III.iv.,110.
14. Reginald Heber, *Narrative of a Journey through the Upper Provinces of India*, London: Murray, 1828, vol. 2, p.330.
15. Stepan P. Krasheninnikov, *The History of Kamtschatka*, tr. James Grieve, Gloucester: Raikes, 1764, pp. 62, 74.
16. Wisdom of Solomon 11:21.
17. *Annals of Philosophy*, 1, London: Robert Baldwin, 1813, pp. 452–7.

Chapter 12   Enlightenment: Leisure, Electricity and Chemistry

1. Samuel Pepys, *The Diary*, ed. Robert Latham and William Matthews, vol. 5, London: Bell, 1971, p. 33 (1 February 1664).
2. B. le B. de Fontenelle, *A Week's Conversation on the Plurality of Worlds*, 6th edn, tr. Mrs A. Behn, Mr J. Glanvill, John Hughes Esq. and William Gardner Esq., London: Bettesworth, 1737, pp. 179–80.
3. Ibid., pp. 183–4.
4. John Heath-Stubbs and Phillips Salman, eds, *Poems of Science*, London: Penguin, 1984, p. 150.
5. David Hume, *The Natural History of Religion* and *Dialogues concerning Natural Religion*, ed. A. Wayne Colver and J.V. Price, Oxford: Oxford University Press, 1976, pp. 221, 241.
6. F.M.A. de Voltaire, *The Elements of Sir Isaac Newton's Philosophy*, tr. John Hanna, London: Austen, 1738, pp. 3, 186.
7. Benjamin Franklin, *Experiments and Observations on Electricity*, ed. I.B. Cohen, Cambridge, MA: Harvard University Press, 1941, p. 180.
8. Joseph Priestley, *The History and Present State of Electricity*, 3rd edn (1775), ed. Robert Schofield, New York: Johnson, 1966, vol. 1, pp. 14–15.
9. Stephen Hales, *Vegetable Staticks* (1727), ed. Michael Hoskin, London: Oldbourne, 1961, pp. 178–9.
10. Joseph Priestley, *Experiments and Observations on Different Kinds of Air*, Birmingham: Pearson, 1790, vol. 2, p. 103.
11. J. Dugald Stewart, *Philosophical Essays,* Edinburgh: Constable, 1810, p. xiii.
12. Priestley, *Experiments . . . on Air*, vol. 1, pp. xix, xx, note 10.
13. Humphry Davy, *Salmonia: or Days of Fly Fishing*, 3rd edn, London: Murray, 1832, pp. 325–6.
14. Kathleen Coburn, ed., *Inquiring Spirit: A New Presentation of Coleridge from his Published and Unpublished Prose Writings*, London: Routledge, 1951, pp. 399, 381.
15. William Wordsworth, *Poetry and Prose*, ed. W.M. Merchant, London: Rupert Hart-Davis, 1955, p. 124.
16. John Keats, *Lamia*, 1820, pt 2, ll. 229, 234–8.
17. Heath-Stubbs and Salman, eds, *Poems of Science*, p. 139.

Chapter 13   Revolution, Evolution: How, Then, Did Science Grow?

1. Albert Einstein, 'Foreword', in Isaac Newton, *Opticks* (1730), New York: Dover, 1952, p. lix.
2. Joseph Priestley, *The History and Present State of Electricity* (1775), New York: Johnson, 1966, pp. xxi, xxii, xxiii.
3. Alfred Tennyson, *Poetical Works*, Oxford: Oxford University Press, 1953, p. 96.
4. Isaiah 11:6, 40:4.
5. Revelation 21:10.
6. Humphry Davy, *Collected Works*, ed. John Davy, London: Smith Elder, 1839, vol. 2, p. 323.

FURTHER READING

There are many history of science societies, both national and global in membership, that hold meetings and conferences; many also publish journals and newsletters. The oldest is the History of Science Society (HSS), based in North America, which publishes *Isis*; the British Society for the History of Science (BSHS) publishes the *British Journal for the History of Science* (*BJHS*); the Society for the History of Alchemy and Chemistry (SHAC) publishes *Ambix*; Newcomen: The International Society for the History of Engineering and Technology and the Society for the Social History of Medicine both publish journals too. All have websites with useful and trustworthy links, and also publish more accessible newsletters. Valuable compilations like the *Dictionary of Scientific Biography*, the *Oxford Dictionary of National Biography*, other national dictionaries, and the correspondence and archives of certain scientists are also available online. The history of science has wide ramifications, and its societies often hold conferences in conjunction with other societies that are concerned with different branches of history – of science, industry, art and collecting, and other topics. Scientific societies also usually have sections dedicated to the history of their discipline. Plunge in!

General Studies

Cohen, H. Floris. *How Modern Science Came into the World: Four Civilizations, One 17th-Century Breakthrough*, University Press, Amsterdam, 2010.

Dear, Peter, ed. *The Literary Structure of Scientific Argument: Historical Studies*, University of Pennsylvania Press, Philadelphia, PA, 1991.

Fara, Patricia. *Science: A Four Thousand Year History*, Oxford University Press, Oxford, 2009.

Feyerabend, P. *Against Method*, Verso, London, 1993.

Galison, Peter, and David J. Stump, eds. *The Disunity of Science: Boundaries, Contexts and Power*, Stanford University Press, Stanford, CA, 1996.

Grant, Edward. *A History of Natural Philosophy, from the Ancient World to the Nineteenth Century*, Cambridge University Press, Cambridge, 2007.

Gregory, Frederick. *Natural Science in Western History*, Houghton Mifflin, Boston, 2008.
Harrison, Peter, Ronald Numbers and Michael Shanks. *Wrestling with Nature: From Omens to Science*, Chicago University Press, Chicago, 2011.
Heilbron, J.L., ed. *The Oxford Companion to the History of Modern Science*, Oxford University Press, Oxford, 2003.
Knight, David. *Public Understanding of Science: A History of Communicating Scientific Ideas*, Routledge, London, 2006.
Ravetz, Jerome. *The No-Nonsense Guide to Science*, New Internationalist, Oxford, 2005.
Shapin, Steven. *The Scientific Revolution*, Chicago University Press, Chicago, IL, 1998.
Stewart, Larry. *The Rise of Public Science: Rhetoric, Technology and Natural Philosophy in Newtonian Britain, 1660–1750*, Cambridge University Press, Cambridge, 1992.

Early Days

Bede. *On the Nature of Things and On Times*, ed. and trans. C.B. Kendall and F. Wallis, Liverpool University Press, Liverpool, 2010.
Bennett, Jim, and Stephen Johnston. *The Geometry of War, 1500–1750*, Museum of the History of Science, Oxford, 1996.
Blackburn, Bonnie, and Leofranc Holford-Strevens, *The Oxford Companion to the Year*, Oxford University Press, Oxford, 1999.
Copenhaver, Brian P., ed. and trans. *Hermetica: The Greek Corpus Hermeticum and the Latin Asclepium*, Cambridge University Press, Cambridge, 1992.
Henderson, George. *Vision and Image in Early Christian England*, Cambridge University Press, Cambridge, 1999.
Hiatt, Alfred. *Terra Incognita: Mapping the Antipodes before 1600*, Chicago University Press, Chicago, IL, 2008.
Jones, Peter Murray, ed. *Medicina Antiqua: Codex Vindobonensis 93*, Harvey Miller, London, 1999.
Po-chia Hsia, R. *A Jesuit in the Forbidden City: Matteo Ricci, 1552–1610*, Oxford University Press, Oxford, 2010.
Scafi, A. *Mapping Paradise*, British Library, London, and Chicago University Press, Chicago, IL, 2006.
Sung Ysing-Hsing. *T'ien-kung K'ai-wu: Chinese Technology in the Seventeenth Century*, Penn State University Press, University Park, PA, 1966.

Chymistry to Chemistry

Abraham, Lyndy. *A Dictionary of Alchemical Imagery*, Cambridge University Press, Cambridge, 1998.
Boantza, Victor D. *Matter and Method in the Long Chemical Revolution: Laws of Another Order*, Ashgate, Aldershot, 2013.
Brock, William H. *The Fontana History of Chemistry*, HarperCollins, London, 1992.
Debus, Allen G. *The Chemical Promise: Experiment and Mysticism in the Chemical Philosophy, 1550–1800*, Science History, Sagamore Beach, MA, 2006.
Debus, Allen G., ed. *Alchemy and Early Modern Chemistry: Papers from Ambix*, Society for the History of Alchemy and Chemistry, London, 2004.
Knight, David. *Ideas in Chemistry: A History of the Science,* Athlone, London, 1992.
Newman, William R. *Gehennical Fire: The Lives of George Starkey, an American Alchemist in the Scientific Revolution*, Harvard University Press, Cambridge, MA, 1994.

Newman, William R. *Promethean Ambitions: Alchemy and the Quest to Perfect Nature*, Chicago University Press, Chicago, IL, 2003.

Nummedal, Tara. *Alchemy and Authority in the Holy Roman Empire*, Chicago University Press, Chicago, IL, 2007.

Pagel, Walter. *John Baptista van Helmont*, Cambridge University Press, Cambridge, 1982.

Shapin, S., and S. Schaffer. *Leviathan and the Air-Pump: Hobbes, Boyle and the Experimental Life*, Princeton University Press, Princeton, NJ, 1985.

Walton, Michael T. *Genesis and the Chemical Philosophy: True Christian Science in the Sixteenth and Seventeenth Centuries*, AMS Press, New York, 2011.

The New Philosophy

Beretta, M., A. Clericuzio and L.M. Principe, eds. *The Accademia del Cimento and its European Context*, Watson, Sagamore Beach, MA, 2009.

Boyle, Robert. *A Free Enquiry into the Vulgarly Received Notion of Nature*, ed. Edward B. Davis and Michael Hunter, Cambridge University Press, Cambridge, 1996.

Clark, W. *Academic Charisma and the Origins of the Research University*, Chicago University Press, Chicago, IL, 2006.

Field, J.V. *Kepler's Geometrical Cosmology*, Athlone Press, London, 1988.

Freedberg, David. *The Eye of the Lynx: Galileo, his Friends and the Beginning of Natural History*, Chicago University Press, Chicago, IL, 2002.

Hotson, Howard. *The Reformation of Common Learning: Post-Ramist Method and the Reception of the New Philosophy, 1618–70*, Oxford-Warburg Studies, Oxford, in press.

Hunter, Michael. *Boyle: Between God and Science*, Yale University Press, New Haven, CT, 2009.

Iliffe, R. *A Very Short Introduction to Newton*, Oxford University Press, Oxford, 2007.

Jardine, Lisa. *On a Grander Scale: The Outstanding Career of Sir Christopher Wren*, HarperCollins, London, 2002.

Knight, Harriet. *Organising Natural Knowledge in the Seventeenth Century*, Lap Lambert, Saarbrücken, 2011.

North, John. *Cosmos: An Illustrated History of Astronomy and Cosmology*, Chicago University Press, Chicago, IL, 2008.

Van Eck, Caroline. *Classical Rhetoric and the Visual Arts in Early Modern Europe*, Cambridge University Press, Cambridge, 2007.

Science and Religion

Brooke, John H., and Geoffrey Cantor. *Reconstructing Nature: The Engagement of Science and Religion*, T. & T. Clarke, Edinburgh, 1998.

Jacob, W.M. *The Clerical Profession in the Long Eighteenth Century, 1680–1840*, Oxford University Press, Oxford, 2007.

Knight, David. *Science and Spirituality: The Volatile Connection*, Routledge, London, 2004.

Knights, Mark. *The Devil in Disguise: Deception, Delusion and Fanaticism in the Early English Enlightenment*, Oxford University Press, Oxford, 2011.

Livingstone, David N. *Adam's Ancestors: Race, Religion and Politics of Human Origins*, Johns Hopkins University Press, Baltimore, MD, 2008.

Manning, Russell Re, ed. *The Oxford Handbook of Natural Theology*, Oxford University Press, Oxford, 2013.

Thomas, Keith. *Religion and the Decline of Magic: Studies in Popular Belief in Sixteenth- and Seventeenth-Century England*, Weidenfeld & Nicolson, London, 1971.

Walsham, Alexandra. *The Reformation of the Landscape: Religion, Identity and Memory in Early Modern Britain and Ireland*, Oxford University Press, Oxford, 2011.
Wood, Paul. *Science and Dissent in England, 1688–1945*, Ashgate, Aldershot, 2004.

Health, Illness and Death

Cressy, David. *Birth, Marriage and Death: Ritual, Religion and the Life Cycle in Tudor and Stuart England*, Oxford University Press, Oxford, 1997.
Macdonald, Michael. *Mystical Bedlam: Madness, Anxiety and Healing in Seventeenth-Century England*, Cambridge University Press, Cambridge, 1981.
Mortimer, Ian. *The Dying and the Doctors: The Medical Revolution in Seventeenth-Century England*, Royal Historical Society and Boydell Press, Woodbridge, Suffolk, 2009.
Porter, Roy. *Flesh in the Age of Reason*, Allen Lane, London, 2003.
Scull, Andrew. *Hysteria: The Biography*, Oxford University Press, Oxford, 2009.
Wrightson, Keith. *Ralph Tailor's Summer: A Scrivener, his City and the Plague*, Yale University Press, New Haven, CT, 2011.

Machines and Instruments

De Nave, F., and L. Voet. *Plantin-Moretus Museum*, Ludion, Ghent and Amsterdam, 2004.
McConnell, Anita. *Jesse Ramsden (1735–1800), London's Leading Scientific Instrument Maker*, Ashgate, Aldershot, 2007.
Maurice, Klaus, and Otto Mayr. *The Clockwork Universe: German Clocks and Automata, 1550–1650*, Neale Watson, New York, 1980.
Miller, D.P. *James Watt, Chemist: Understanding the Origins of the Steam Age*, Pickering & Chatto, London, 2009.
Morrison-Low, A.D. *Making Scientific Instruments in the Industrial Revolution*, Ashgate, Aldershot, 2007.
Moxon, Joseph. *Mechanick Exercises on the Whole Art of Printing (1683–4)*, ed. H. Davis and H. Carter, 2nd edn, Oxford University Press, Oxford, 1962.
Moxon, Joseph. *Mechanick Exercises, or The Doctrine of Handy-Works*, ed. B.M. Forman, Praeger, New York, 1970.
Suarez, Michael F., and H.R. Woudhuysen, eds. *The Oxford Companion to the Book*, Oxford University Press, Oxford, 2010.
Wigelsworth, Jeffrey R. *Selling Science in the Age of Newton*, Ashgate, Aldershot, 2010.

Natural History

Allen, David E. *The Naturalist in Britain: A Social History*, Princeton University Press, Princeton, NJ, 1994.
Allen, David E. *Books and Naturalists*, HarperCollins, London, 2010.
Baldner, Leonhard. *Vogel-, Fisch- und Thierbuch*, ed. R. Lauterborn, Verlag Müller und Schindler, Stuttgart, 1973.
Fisher, Celia. *Flowers of the Renaissance*, Frances Lincoln, London, 2011.
Gage, Andrew Thomas, and William Thomas Stearn. *A Bicentenary History of the Linnean Society of London*, Academic Press, London, 1988.
Graham, Michael H., Joan Parker and Paul K. Dayton. *The Essential Naturalist: Timeless Readings in Natural History*, Chicago University Press, Chicago, IL, 2011.

Lister, Martin. *English Spiders*, trans. M. Davies and B. Hartley, ed. J. Parker and B. Harley, Harley Books, Colchester, 1992.

Nelson, Charles, ed. *History and Mystery: Notes and Queries from the Newsletters of the Society for the History of Natural History*, SHNH, London, 2011.

Roos, Anna Marie. *Web of Nature: Martin Lister (1639–1712), the First Arachnologist*. Brill, Leiden, 2011.

Stafleu, Frans A. *Linnaeus and the Linneans: The Spreading of their Ideas in Systematic Botany, 1735–1789*, A. Oosthoek's Uitgeversmaatschappij, Utrecht, 1971.

Exploring and Mapping

Beaglehole, J.C., ed. *The Journals of Captain James Cook*, Hakluyt Society, Cambridge, 1955–69.

Beaglehole, J.C., ed. *The Endeavour Journal of Joseph Banks*, Angus & Robertson, and Public Library of NSW, London, 1962.

Carr, D.J., ed. *Sydney Parkinson: Artist of Cook's Endeavour Voyage*, Croom Helm and British Museum (Natural History), London, 1983.

Cumming, W.P., R.A. Skelton and D.B. Quinn. *The Discovery of North America*, Paul Elek, London, 1971.

Cumming, W.P., S.E. Hillier, D.B. Quinn and G. Williams. *The Exploration of North America, 1630–1776*, Paul Elek, London, 1974.

David, Andrew, ed. *The Charts and Coastal Views of Captain Cook's Voyages*, vol. 1, Hakluyt Society, London, 1988.

Harvey, Paul D.A. *The History of Topographical Maps: Symbols, Pictures and Surveys*, Thames and Hudson, London, 1980.

Joppien, Rüdiger, and Bernard Smith, *The Art of Captain Cook's Voyages*, Yale University Press, New Haven, CT, 1985.

McLynn, Frank. *Captain Cook: Master of the Seas*, Yale University Press, New Haven, CT, 2011.

Pedley, Mary Sponberg. *The Commerce of Cartography: Making and Marketing Maps in Eighteenth-Century France and England*, Chicago University Press, Chicago, IL, 2005.

Terrall, Mary. *The Man Who Flattened the Earth: Maupertuis and the Sciences in the Enlightenment*, Chicago University Press, Chicago, IL, 2002.

The Enlightenment

Anderson, Robert G.W., and Jean Jones, *The Correspondence of Joseph Black*, Ashgate, Aldershot, 2012.

Eddy, Matthew D. *The Language of Mineralogy: John Walker, Chemistry and the Edinburgh Medical School*, Ashgate, Aldershot, 2008.

Fara, Patricia. *Erasmus Darwin: Sex, Science and Serendipity*, Oxford University Press, Oxford, 2012.

Langford, Paul. *A Polite and Commercial People: England 1727–1783*, Oxford University Press, Oxford, 1989.

Uglow, Jenny. *The Lunar Men: The Friends Who Made the Future*, Faber, London, 2002.

Into Modern Times

Crosland, Maurice. *Science under Suspicion*, Guildford, Grosvenor, 2011.

Holmes, Richard. *The Age of Wonder: How the Romantic Generation Discovered the Beauty and Terror of Science*, Harper Press, London, 2008.

Hughes-Hallett, Penelope. *The Immortal Dinner: A Famous Evening of Genius and Laughter in Literary London, 1817*, Vintage, London, 2012.

Knight, David. *The Making of Modern Science: Science, Technology, Medicine and Modernity, 1789–1914*, Polity, Cambridge, 2009.

Schrempp, Gregory. *The Ancient Mythology of Modern Science: A Mythologist Looks (Seriously) at Popular Science Writing*, MacGill-Queen's University Press, Montreal, 2012.

Shapin, Steven. *The Scientific Life: A Moral History of a Late Modern Vocation*, Chicago University Press, Chicago, IL, 2008.

Illustration Acknowledgements

All books from which the illustrations are taken (except number 28, which comes from a book belonging to the author) are reproduced by kind permission of Durham University Library. Their shelf marks are:

1. GAAR.A90A; 2. S.R.2.C.2; 3. Bamburgh 1.1.1-5; 4. Cosin W.5.32; 5. S.R.2.B.4; 6. Bamburgh Select L.3.15; 7. SB 0106; 8. BABC.B84A; 9. Bamburgh Select F.IV.7; 10. SC+00727; 11. SB 0255; 12. Bamburgh C.4.35-36; 13. Cosin Y.1.16; 14. SB+ 00003238; 15. SC+ 00043; 16. Bamburgh M2 6-7; 17. SC+ 00034; 18. Bamburgh 1.3.3; 19. Bamburgh Q.111.6; 20. Bamburgh Select N.3.19; 21. Bamburgh P.6.1; 22. Bamburgh P.3.1; 23. Bamburgh 1.5.40; 24. Bamburgh 1.V.40; 25. SB+ 0009; 26. Bamburgh F.3.49; 27. Bamburgh L.111.53; 29. Routh 51.K.18/13; 30. Cosin T.1.6; 31. Kellet 309; 32. XX598.2BEL; 33. Bamburgh L.VI.61; 34. SC+ 00092; 35. SC+ 00093; 36. Cosin T.2.1-3; 37. Routh 69.C.23; 38. Bamburgh N.1.27; 39. Cosin T.11.15; 40. Bamburgh F.111.46; 41. Elliott 1.10; 42. Bamburgh L.6.4; 43. +ELCN.C84B; 44. Routh 57.C.8; 45. Routh 59A7; 46. SC+ 00198; 47. SC+ 00334; 48. Routh 53.B.10; 49. SC+ 00066; 50. SC 00335; 51. SC 01020; 52. SC+ 00460; 53. Winterbottom +J9/1; 54. SC 00843-00844; 55. Winterbottom +J17.

ACKNOWLEDGEMENTS

I would like to thank my colleagues in the Philosophy Department at Durham University for their forbearance and support in making retirement a long-drawn-out process rather than an event; and for invitations from time to time to present work in progress to the weekly Research Seminar. Thanks especially also to Sheila Hingley and her staff in the Rare Books collection in Durham University Library, particularly Richard Higgins, who patiently assembled the books from which the illustrations are taken, and Caroline Craggs, who scanned them.

Many thanks to Heather McCallum, my editor at Yale University Press, who persuaded me to write this book, to her colleague Rachel Lonsdale for her valuable comments on the draft, and to Michael Hunter and Trevor Levere who have preserved me from sundry errors, omissions and misconceptions; for those that survive I am sorry.

I am very grateful to colleagues in the British Society for the History of Science, the Society for the History of Alchemy and Chemistry, the Royal Institution, and many other places at home and abroad that have continued to invite me to listen and to hold forth, and thus educate myself in public; and to the manuscript reviewers who have persuaded me to read carefully all sorts of books that I might otherwise have skimmed or missed.

Thank you also to the students at Durham University, who over the years have come to my lectures on history of science, asked awkward questions, and written and presented essays and dissertations, all of which activity has forced me to think – and, I hope, express myself – more clearly.

INDEX